Handbook of Nonlinear Optics

Handbook of Nonlinear Optics

Edited by **Kristie Ames**

LANRYE
INTERNATIONAL

New Jersey

Published by Clanrye International,
55 Van Reypen Street,
Jersey City, NJ 07306, USA
www.clanryeinternational.com

Handbook of Nonlinear Optics
Edited by Kristie Ames

© 2015 Clanrye International

International Standard Book Number: 978-1-63240-281-3 (Hardback)

Contents

Preface

Major aspects of nonlinear optics have been discussed in this book. Characterization of properties of light traversal in non-linear media has always intrigued scientists and researchers. Accelerated advancement of laser techniques and optoelectronic devices account for a crucial task of formulating and analyzing the structures capable of efficiently transforming, modulating, and recording optical data in a wide spectrum of radiation energy densities and frequencies at one hand, and novel approaches and schemes capable of activating and stimulating the contemporary features on the other. It is a known fact that the phenomena and materials of nonlinear optics have an assuring place in dealing with these intricate technical tasks. The state-of-the-art ideas, approaches, and information presented in this book will be beneficial to the readers in developing a sustainable solution in basic analysis as well as industrial approach. The aim of this book is to serve as a useful source of information for students, researchers, engineers and technical officers of optoelectronic universities and companies.

After months of intensive research and writing, this book is the end result of all who devoted their time and efforts in the initiation and progress of this book. It will surely be a source of reference in enhancing the required knowledge of the new developments in the area. During the course of developing this book, certain measures such as accuracy, authenticity and research focused analytical studies were given preference in order to produce a comprehensive book in the area of study.

This book would not have been possible without the efforts of the authors and the publisher. I extend my sincere thanks to them. Secondly, I express my gratitude to my family and well-wishers. And most importantly, I thank my students for constantly expressing their willingness and curiosity in enhancing their knowledge in the field, which encourages me to take up further research projects for the advancement of the area.

Editor

Part 1

Features of Nonlinear Optics Mechanisms

Overview of Nonlinear Optics

Elsa Garmire
Dartmouth College,
USA

1. Introduction

The invention of the laser provided enough light intensity that nonlinear optics (NLO) could be observed for the first time, almost exactly one year after the first ruby laser – second harmonic generation, observed fifty years ago (Franken, 1961), with a theoretical examination of interactions between light waves in a nonlinear dielectric following very soon thereafter (Armstrong, 1962). The field has grown so enormously that it is impossible to review all topics. Theoretical approaches have bifurcated into macroscopic and microscopic viewpoints. Both have validity and usefulness, but this review will focus primarily on experimental results and applications, as well as models for optical nonlinearities that stress the macroscopic character of materials, with parameters determined experimentally. At times the wavelike nature of light is sufficient to explain concepts and at other times it is easier to refer to the photon-like nature of light. In nonlinear optics we become fluent in both concepts.

Table 1 shows a classification scheme that tries to put some order into the many phenomena that comprise nonlinear optics. Optical nonlinearities occur when the output of a material or device ceases to be a linear function of the input power, which is almost always the case for high enough intensities. The nonlinearity may cause a light-induced change in refractive index or absorption of the medium or it may cause new frequencies to be generated.

Character of medium:	Transparent	Absorbing	Scattering	Non-Local	
Atom-light interaction:	Non-resonant	Resonant	Incoherent	Coherent	Transient
Optics geometry:	Plane Wave	Finite Beam	Waveguide	Wave-Mixing	Reflection
Device Geometries:	Bulk	Periodic	Fibers	Resonators	Micro-cavities

Table 1. Classification of nonlinear optical phenomena

1.1 Character of the medium

The character of the dielectric medium may be such that it is transparent, absorbing or scattering to the incident light. Usually the medium responds to the local optical intensity;

however, in some cases the nonlinearity is non-local, so that light intensity at one point creates a change in absorption or refractive index at another point. Examples of non-local nonlinearities are absorptive thermal effects and optically-induced carrier-transport, such as occurs in photo-refractivity.

1.2 Atom-light interaction

When its frequency does not match any atomic or molecular resonance, low intensity light is transmitted through transparent media without loss; this non-resonant interaction with the medium is expressed as a refractive index. At higher intensities, the nonlinearity is due to a nonlinear refractive index. Stimulated scattering such as Stimulated Raman Scattering (SRS), and Stimulated Brillouin Scattering (SBS) phenomena represent another class of non-resonant nonlinearities.

When its frequency is resonant with atomic or molecular transitions, light is absorbed, with a magnitude expressed in terms of an absorption coefficient. At high intensity, saturable absorption may occur, due to filling the available upper states. In this case the absorption begins to decrease with intensity and the material may become transparent. Even if the medium is initially transparent, when the incident light is sufficiently intense, multi-photon absorption may take place, as a result of near-resonance between the medium's absorption lines and multiples of the incident light frequency. Multi-photon absorption causes an increasingly large absorption as the light intensity increases.

Typically individual atoms or molecules lose their coherence during their interaction with light, so that the light-matter effects can be understood as incoherent phenomena. However, with sufficiently short pulses, or long coherence times, nonlinear phenomena will demonstrate coherent interactions between the light and the medium, in which phase is preserved. Examples are self-induced transparency and pi pulses. A number of transient phenomena do not require phase coherence – an example is self-phase modulation.

Nonlinear optics occurs when any of these interactions are changed by the intensity of the incident light, thereby affecting the output. Some authors consider "nonlinear optics" to be any phenomenon that changes the way light interacts with a medium, such as by changing the refractive index through applying an external voltage (e.g. electro-optical modulators). This chapter does *not* consider these effects. The formal definition of nonlinear optics used here requires that the light itself must cause the material properties to change and thereby change how it interacts with light (either the initial lightwave or another lightwave).

In almost all of its manifestations, nonlinear optics can be explained classically – quantum mechanics is not usually required. Most of the nonlinear effects can be explained by a macroscopic view of the medium through a nonlinear polarization and/or nonlinear absorption coefficient. In some cases it is more useful to consider how the microscopic behavior of individual molecules adds up to result in nonlinear phenomenon – this is particularly true with resonant interactions.

1.3 Optics geometries

Many optical nonlinearities use a broad beam of light incident on a nonlinear medium, which is modeled as a plane wave. Direct observation of harmonics generated from a plane wave can occur as long as there is *phase-matching*. Direct observation of nonlinear absorption is made through intensity measurements. The nonlinear refractive index is often observed through changes in the spatial profile of a beam of finite width, which is most dramatic with self-focusing or thermal blooming and is described in the Z-scan discussion in section 2.

Because diffraction from a finite beam shortens the length over which the nonlinearity is large, waveguides are often used in nonlinear optics geometries. If the phase-matching condition is met, the beam can be focused into a waveguide and won't spread out; the nonlinear interaction can remain large over rather long lengths.

Wave-mixing describes the fact that two plane waves overlapping within a nonlinear medium can mix the waves together in a nonlinear way. This may occur from nonlinear refractive index and/or nonlinear absorption. The nonlinearity introduces a periodicity in the medium due to these mixing waves that refracts additional waves. The result is a nonlinear diffracting grating. There can be two-wave mixing, three-wave mixing or four-wave mixing. Measuring the nonlinear refractive index by interfering a beam with a reference beam through a nonlinear medium represents another kind of wave-mixing.

Finally, optical nonlinearities can occur in reflection from a nonlinear interface, which can be demonstrated using prism coupling. Surfaces can enhance optical nonlinearities, particularly in micro-cavities.

1.4 Device geometries

All of the nonlinear processes described above can be used in a variety of device geometries. Descriptions generally begin with bulk media, but can be replaced by periodic media that can be quasi-phase-matched, or that provide reflective gratings. Fibers have the advantage of producing intense light in waveguides of very small cores over very long distances. While the glass fibers usually have small nonlinearities, most nonlinear phenomena can build up to be quite large in fibers because of their very long length.

Optical resonators can enhance the optical cavity field, effectively enlarging the optical nonlinearities. An example is the intra-cavity frequency-doubled diode-pumped solid-state laser used today as the ubiquitous green laser pointer. The Fabry-Perot etalon was shown to demonstrate optical bistability, when filled with a nonlinear refractive index and/or absorpting material. Another form of optical resonator is the micro-sphere (even quantum dots), which have exhibited enhanced optical fields and therefore enhanced nonlinearities. Closely related, but not detailed here, are enhanced nonlinearities due to surface polaritons at metal-dielectric surfaces.

1.5 NLO and its applications

Nonlinear optics is a very broad field, centered in both Physics and Electrical Engineering, more specifically in the sub-fields of Optics and Photonics. Select the topic "nonlinear optics" in the ISI Web of Science and I found 7,240 papers in September, 2011. Google found 910,000 pages on the same day! It is interesting that the 50 most cited papers were published in 26 different journals! This is because nonlinear optics impacts a wide range of technical fields, including optical communications, fiber optics, ultrafast lasers, quantum computing, ultra-cold atoms, plasma physics, particle accelerators, etc. Both chemistry and biology are using increasing amounts of nonlinear optics. Nonlinear optics applies to numerous specific applications, such as in fiber optics, spectroscopy, photorefractivity, liquid crystals, polymers, semiconductors, organics, switching, ultraviolet, X-rays, quantum optics, telecommunications and signal processing. This review will not explore applications in detail, only introduce a few of the most important. A great deal of information about nonlinear optics is now available on the web.[1]

This chapter will introduce the important basic phenomena of nonlinear optics and their applications, along with some of the most important concepts. Macroscopic, classical concepts will be emphasized, although simple quantum mechanics concepts can be introduced in a straight-forward manner.

2. Nonlinear polarization density

The origin of nonlinear optics is the nonlinear response of the material to internal electric fields. The material response is manifest in a *susceptibility*. The linear susceptibility is related to the refractive index through $\chi_1 = n^2 - 1$. When intense light enters a transparent material, its susceptibility can become nonlinear. It is usually sufficient to consider this heuristically as a Taylor expansion in the electric field:

$$\chi(E) = \chi_1 + \chi_2 E + \chi_3 E^2 + \chi_4 E^3 \dots \tag{1}$$

All higher orders beyond χ_1 represent the nonlinear response of a material to the presence of intense light. The higher order terms fall off in magnitude very rapidly unless the light intensity is very high. In general the susceptibility is a tensor and the vector component of the fields must be taken into account.

The effect of the susceptibility on the light traveling through the medium is manifest in the *polarization density,* **P**. In tensor notation the linear polarization is $\mathbf{P} = \varepsilon_0 \chi_1 : \mathbf{E}$. The nonlinear polarization density is $\mathbf{P}_{NL} = \varepsilon_0 \chi(E) : \mathbf{E}$, so the second order polarization is $\mathbf{P}_2 = \varepsilon_0 \chi_2 : \mathbf{EE}$ and the third-order nonlinear polarization is $\mathbf{P}_3 = \varepsilon_0 \chi_3 : \mathbf{EEE}$. There is no universal notation, however; the nonlinear refractive index (a third-order nonlinearity) is often written in terms of intensity as $n(I) = n_0 + n_2 I$. Thus the nonlinear component of the refractive index that is linear in intensity is called n_2, while the related susceptibility is χ_3.

2.1 Second-order nonlinear polarization density, P_2.

For two incident waves, one at frequency ω_0, and the other at ω_1, the nonlinear term χ_2 will introduce (ignoring tensor notation) $P_2 = \chi_2 E_0 \cos(\omega_0 t - k_0 z) E_1 \cos(\omega_1 t - k_1 z)]$, where amplitudes are E_0 and E_1 and wave vectors k_0 and k_1 are related to frequencies by their respective velocities of light. This product gives two polarization terms, one that oscillates at $\omega_0 + \omega_1$, and the other at $\omega_0 - \omega_1$. Both terms are proportional to the product of the fields. From the quantum mechanical point of view, the nonlinearity has induced two photons to combine into one photon. When the two photons have the same frequency, one term yields *second*

---

[1] Wikipedia is quite reliable for investigating the phenomena outlined in this paper. Almost any of the topics discussed in this paper will be found in Wikipedia. See also, "An Open Access Encyclopedia for Photonics and Laser Technology," written by Dr. Rüdiger Paschotta, a consultant in photonics, http://www.rp-photonics.com/topics_nonlinear.html, which covers the basics of nonlinear optics rather well, with good search capability.

harmonic and the other yields a term with a *static field* (the frequency dependence cancels out). When the incident photons are different, *sum* and *difference* frequency photons are generated. An important criterion for materials to exhibit a susceptibility linear in the field, X_2E, is that they contain no center of inversion symmetry. Liquids, gases, amorphous solids and crystalline materials with high symmetry will not directly generate second harmonic. Typically the goal of X_2E terms is to transfer power from one frequency to another, while maintaining a coherent beam. This requires *phase-matching*, which will be discussed later.

2.1.1 Second Harmonic Generation (SHG)

Perhaps the most important application of P_2 is converting infrared light into visible. The green laser pointer consists of a diode-pumped solid state laser emitting in the infrared at 1.06 µm that is frequency-doubled by a nonlinear crystal to a wavelength of 0.53 nm. For one incident wave of frequency ω_o, the nonlinear term χ_2 introduces into the polarization P_2 a term that oscillates at $2\omega_o$, the second harmonic. Applications for SHG go considerably beyond laser pointers. Lasers that directly emit visible light are less efficient than infrared lasers, so when visible light is required, it is preferable to start with the more efficient infrared lasers and to frequency-double them. SHG has been a standard complement for Nd:YAG lasers for a long time. Diode-pumping has replaced lamp-pumping for most of these applications, increasing their efficiency. Visible (or ultra-violet) lasers are commonly used to pump other lasers, most notably the titanium-sapphire laser (and formerly, the dye laser). The highly inefficient argon laser is rapidly being replaced by frequency-doubled diode-pumped solid state lasers for applications such as pumping the ultra-short pulse titanium-sapphire lasers, which can be mode-locked to pulses only a few femtoseconds long and emit at 800-900 nm wavelengths.

How are these femtosecond pulses measured? With an *autocorrelator* that measures the physical length of fs pulses by means of SHG. An autocorrelator is created by splitting an ultra-short pulse train into two beams, and colliding them in a SHG crystal in such as way that the harmonic occurs only when both pulses overlap (a particular crystal geometry). Delaying one beam with respect to the other, and scanning this delay, enables the length of the pulse in physical space to be determined. Without SHG, the entire field of ultra-fast optics would be severely hampered.

In addition to simply offering light you can see, SHG is important because each visible or UV photon has enough energy to cause a chemical reaction. In non-homogenous materials, the generation of second harmonic may select for specific regions. Examples range from separating out collagen and microtubules in live tissue (Zipfel, 2003) to observing coupled magnetic and electric domains in ferroelectromagnets (Fiebig, 2003). Surface science is an important application because the surface breaks the symmetry of the bulk and enables SHG that depends critically on the character of the surface. The surface can also offer resonance enhancement of the signal (Hsu, 2011); monolayer adsorption can be detected, for example of tin on GaAs (Shen, 1994; Mitchell, 2009). In other applications, surface SHG can monitor laser melting and separate amorphous from crystalline growth. As a spectroscopic tool, SHG has been used in a plethora of applications, such as probing surface states of metals, surface magnetization, and, using ultrashort pulses, a wide range ultrafast surface reactions and surface dynamics.

2.1.2 Sum and Difference Frequency Generation (SFG and DFG)

When the incident field E contains two frequencies ω_o and ω_i, sum frequency and/or difference frequency generation is possible, as seen directly from Eq. (1); the cross-term gives a polarization of the form $P_2 = \chi_2 E_o E_i$. SFG can convert infrared light at frequency ω_i into a visible signal at frequency $\omega_s = \omega_o + \omega_i$. When light at frequency ω_o is very intense, there is even an effective amplification of the weak infrared signal. SFG is one way to provide coherent UV light from visible light. If one of the visible lasers has a tunable frequency, the UV light's frequency can be tuned.

DFG has been used to create infrared light from two higher frequency laser beams. The term DFG usually refers to the case where the beams at the two incident frequencies have comparable intensity. The output infrared signal will be at $\omega_s = \omega_o - \omega_i$. When one beam is very intense and the other is weak, amplification will occur (as with SHG). This is often called *parametric amplification*, which will be described later.

DFG has applications in telecommunications, where Wavelength Division Multiplexing (WDM) puts many wavelengths on the same optical fiber. In real WDM systems, a way is needed to convert from one wavelength to another. DFG is attractive in several respects: it is an instantaneous process that can simultaneously convert up and down multiple channels with equal efficiencies, has negligible spontaneous emission noise and no intrinsic frequency chirp (Yoo, 1996; Yu, 2007).

2.1.3 Nonlinear Optical Rectification (NOR)

The nonlinear term corresponding to $P_r = \varepsilon_0 \chi_2 E E^*$ has no time dependence, other than that of the magnitude of E. It creates a DC static polarization through the time-average of E^2. This process is traditionally called *optical rectification*. The major practical application of optical rectification has been the generation of Terahertz radiation through optical rectification of femtosecond pulses (Dragoman, 2004; Fueloep, 2011). Because of the time-dependence of these fs pulses, the "static" field is not time-independent, but rises and falls within the width of the fs pulse, generating broadband electromagnetic pulses in free space. This results in fields with variation at Terahertz frequencies waves that have sub-mm wavelengths. Coherent light in the THz frequency domain is rather new and applications are presently being developed, predominantly spectroscopy. For optical rectification, the χ_2 materials ZnTe and GaAs are typically used.

Other applications for NOR include accelerating electrons within a small distance by means of static field enhancement due to surface plasmons. This enhancement may provide nanoscale geometries for high-energy electron sources, where electrons are accelerated in the electric field of surface plasmons (Lenner, 2011). Just as SHG is sensitive to surface conditions, so is NOR. It has become an important technique for investigating semiconductor heterostructures and nanostructures, such as asymmetric quantum wells (Wang, 2009). Optical rectification has even enabled miniature silicon waveguides to detect light within the bandgap that would not normally be detected (Baehr-Jones, 2005).

One of the most exciting applications of NOR comes in "wakefield generation," which is a relativistic version of optical rectification that introduces longitudinal field effects that can be as large as transverse effects. Electromagnetic field intensities in excess of 10^{18} W/cm^2 lead to relativistic electron motion in the laser field. In addition to NOR, other effects may occur, including relativistic focusing, relativistic transparency, nonlinear modulation and multiple harmonic generation, and strong coupling to matter and other fields (such as high-frequency radiation) (Mourou, 2006). Optics in the relativistic regime is an exciting new direction for nonlinear optics (Tsaur, 2011).

2.2 Phase-matching

Efficient growth of second harmonic, sum, or difference frequency requires the nonlinear polarization P_2 to remain in the same relative phase with the incident light over the entire interaction length. This is called *phase-matching*, with the phases of the incident and resultant waves remaining in synchronism. Lack of synchronism comes about because the refractive index is a function of wavelength (dispersion), with shorter wavelengths (the harmonics) typically having higher refractive indices than the fundamental. Phase coherence requires $(n_{2\omega} - n_\omega)L < \lambda/4$, which may typically be tens of micrometers. But in SHG the path length required to build up substantial signal is usually the order of centimeters. Thus special techniques are required to match phases, including anisotropic crystals, periodic domain reversal (quasi-phase matching) and waveguides (matching dispersion or using Cerenkov output).

2.2.1 Phase-matching in anisotropic crystals

In *anisotropic crystals* the refractive index depends on the direction of the light's polarization and its propagation direction with respect to the crystal. This dependence enables nonlinear crystals cut in particular geometries to have their fundamental and second harmonic with the same phase velocity. This is the only way that phase-match can be achieved in uniform bulk crystals. Anisotropic crystal phase matching can be very sensitive to crystal angle and temperature; calculations of suitable angles require complicated tensor calculations, but fortunately the commercial venders know how to do this and sell crystals cut to the proper orientation. In *type I phase matching* the fundamental has a single linear polarization and the harmonic has a polarization perpendicular to the fundamental. These crystals can be tuned by changing temperature or by changing the angle, or both. *Type II phase matching* requires that the nonlinear process mix laser beams with orthogonal polarizations. The harmonic can be along either polarization. Critical phase matching uses the angular dependence of the refractive index with beams propagating off-axis to match polarizations, while non-critical phase matching heats the crystal to achieve phase- matching for beams that propagate along the axis (Paschotta, 2011).

Efficient harmonic generation requires highly nonlinear crystals; a great deal of research has gone into finding and growing suitable material. Originally only crystal quartz and potassium di-hydrogen phosphate (KDP) were available. Lithium niobate was the first of many crystals grown for the purpose of harmonic generation; now a plethora of crystals are available, depending on the particular application and wavelength.

Sum and difference frequency generation is calculated through the coupling between three waves, rather than two, but phase-matching is required for these processes, just as for second harmonic generation.

No one crystal is ideal. Phosphates provide the largest nonlinear crystals: KDP (potassium dihydrogen phosphate), KD*P (deuterated potassium dihydrogen phosphate) and ADP (ammonium dihydrogen phosphate) are all grown in water-solution to very large size, and are quite impervious to optical damage, although hygroscopic. KDP is being grown in tremendous sizes for the inertial confinement fusion laser at Livermore Laboratories in California. The phosphates have transparency to below 200 nm and out to 2 μm, but their nonlinear coefficient is quite small: $d_{eff} \sim 0.4$ pm/V (pm = picometers = 10^{-12}m). By comparison, KTP (potassium titanyl phosphate) and KTA (potassium titanyl arsenate) have an order of magnitude higher nonlinearity: $d_{eff} \sim 4$ pm/V and are non-hygroscopic, but they are hard to grow and only small crystals are available. KTP is commonly used today to frequency-double diode-pumped solid state lasers. However, if the optical power becomes too high, KTP shows photo-induced "gray-tracks."

Lithium niobate has large nonlinearities with a transparency range from 420 nm out to 5.2 μm and a very high nonlinearity (d_{eff} = 5.8 pm/V) but exhibits debilitating photo-refractive effects (see later) unless heated or doped with oxides. Magnesium oxide doping is commonly used. Zinc oxide doping increases the nonlinearity to 16 pm/V and has better optical quality and lower absorption, but is limited to 250 MW/cm^2 of light intensity. Eighty times improvement in damage threshold is obtained with zirconium oxide doping, at the expense of roughly half the nonlinearity. Lithium Tantalate (LiTaO$_3$) has less optical damage than LN and doesn't require doping, but its nonlinearity is half that of LN. Potassium niobate (KNbO$_3$) has high nonlinearity and high damage threshold, which sounds ideal, but it has only a small temperature range of phase-matching, and, more importantly, shaking or stressing the crystal can cause domain changes, so it must be handled "smoothly," limiting its use.

Borates form another class of nonlinear crystals, transparent into the UV, with a high threshold to optical damage, but with considerably lower nonlinearity: LBO (lithium triborate) is transparent to 150 nm, but $d_{eff} \sim 0.8$ pm/V, roughly twice KDP. BBO (beta-barium borate) has a more respectable $d_{eff} \sim 2$ pm/V. Bismuth Borate (BiBO) has $d_{eff} \sim 3.3$ pm/V and the highest damage threshold. These crystals are excellent for high-power applications: they're inert to moisture, with a large acceptance angle and small walk-off angle and suitable for temperature-controllable non-critical phase-matching.

Research is underway on organic crystals for frequency doubling, but they have not yet proven more reliable than the above-mentioned crystals.

In the infrared, a different set of crystals has been developed, with higher nonlinearities but much lower damage thresholds. These include AGS (AgGaS$_2$), transparent over 0.5 - 13 μm; AGSe (AgGaSe$_2$) transparent over 0.7 - 18 μm; also HGS (HgGa$_2$S$_4$) and ZnGeP$_2$. Gallium selenide (GaSe) has the record nonlinearitiy, d_{eff} = 70 pm/V and doubles out to 13 μm wavelength.

The crystals just discussed are anisotropic, required for birefringent phase matching. Periodic poling can provide quasi-phase matching, which removes the need for birefringence. It can also increase the nonlinearity in birefringent crystals; periodic poling of lithium niobate (PPLN) exhibits the effective nonlinear coefficient 4.5 times relative to homogeneous LN.

2.2.2 Quasi-Phase Matching (QPM)

In quasi-phase matching, the nonlinear coefficient is reversed every coherence length (coherence length L_c is the length over which the accumulated phase mismatch = π). Quasi-phase matching is important when birefringent phase-matching is difficult; the crystal need not even be birefringent. Without QPM, the fundamental and harmonic would get out of phase after a distance L_c given by $(k_2 - 2k_1)L = \pm\pi$. If the optical axis is flipped periodically at this distance, the nonlinearity reverses sign and harmonic conversion can continue to build up. Although quasi-phase matching was suggested in the early days of NLO, it did not become practical until the invention of periodic poling (Byer, 1997).

Periodic poling engineers the ferroelectric domains within lithium niobate and similar nonlinear crystals. A strong electric field is applied via patterned electrodes on the crystal surface. Domain reversal occurs at field strengths above the coercive level, which can be anywhere from 2 - 21 kV/mm in LN, depending on the crystal characteristics. The period of the electrode pattern determines the wavelengths for which QPM will occur. This process works best for waveguides and samples less than 0.5 mm thick; it works best in the infrared, where electrode periods can be 10 µm or more. The most common materials that are periodically poled are MgO-doped lithium niobate (called PPLN), lithium tantalate and KTP, while PP potassium niobate has also been reported. Organic nonlinearities such as nonlinear polymers can be poled for QPM (Hung, 2008). Quasi-phase matching has also been reported in glasses and fibers. Its most practical application is often in waveguides.

Periodic poling has extended SHG from anisotropic crystals to polymers and semiconductors. In addition to spatially periodic poling by pulsed electric fields that creates domains with flipped crystal axes, electron bombardment and thermal pulsing have also been successful with QPM. Because of poling enhancement of X_2, harmonic output in PPLN can be many times what would be observed in a single crystal, for the same intensity (Parameswaran, 2002). Second-harmonic generation, difference-frequency generation, and optical parametric oscillation all have used QPM.

2.2.3 Phase-matching in waveguides

The first reason to use waveguides for SHG is that the intensity of the incident light can be maintained over a long length. Before QPM, it was necessary to match the phases of the guided fundamental wave with the guided harmonic. This ability was demonstrated by choosing a wavelength that matched the phases of an incident TM_0 fundamental and the TE_1 mode of its harmonic (Sohler, 1978). Phase-matching in waveguides requires an ability to tailor the guide's effective refractive index. With proper design, the fundamental can create a harmonic in a mode whose effective refractive index will match that of the fundamental, so that waveguide dispersion cancels the material dispersion. The two modes must also have sufficient overlap that a strong harmonic can build up. Some examples have been nonlinear Langmuir-Blodgett films (Penner, 1994), corona-poled sputtered glass films (Okada, 1992), dye polymers (Sugihara, 1991) and complex semiconductor waveguide structures (Abolghasem, 2009; Fiore, 1998; Malis, 2004). To date these techniques have not been reliable enough to be used in practical systems. Often coupling into waveguides is not easy; bulk SHG crystals are in much greater demand. By the year 2000 SHG in waveguides usually has incorporated quasi-phase-matching.

A different approach uses photonic crystal waveguides, which can be tuned to achieve phase-matching for SHG (Martorell, 1997; Broderick, 2000; Mondia, 2003; Torres 2004).

For completeness we mention SHG observed as a Cerenkov-phase-matched harmonic from a waveguide's fundamental mode. The harmonic wave has a phase velocity beyond cutoff for the waveguide, so that it propagates freely in the substrate material, acting as the source of the observed Cerenkov radiation. This phase-matching technique creates a beam at a small angle to the waveguide and has been observed in lithium niobate (Li, 1990; Wang, 1995) as well as nonlinear photonic crystal waveguides (Zhang, 2008). Cerenkov SHG can be important in applications where guided SHG is not desired.

2.3 Theory of second harmonic generation

When there is negligible depletion of the fundamental, the amount of second harmonic is conveniently solved through the coupled mode equations. These equations demonstrate the coupling between the fundamental and its harmonic through the nonlinear susceptibility χ_2. Using complex notation, the polarization density vector introduced by interaction between two input fields is $P_k(\omega_3) = \Sigma_{ij}\, \varepsilon_0(\chi_{ijk}/2)E_i(\omega_1)*E_j(\omega_2)$, where it is important to acknowledge the tensor character of χ_{ijk} For second harmonic, $\omega_1 = \omega_2$ and usually the direction i and j are parallel, so $E_i = E_j$. For any particular configuration of input fields, the summation over all relevant values of χ can be represented by $P_3(2\omega) = d_{eff}\varepsilon_0 E(\omega)^2$. By convention the second harmonic coefficient d_{eff} is most often used as the nonlinear parameter; it is the effective magnitude of the tensor $d_{ijk} = \chi_{ijk}/2$, which depends on the specific configuration of the optical fields.

In lossless materials, the fundamental field E_ω and its harmonic $E_{2\omega}$ can be related by the following coupled mode equations:

Fundamental:

$$\partial E_\omega/\partial z \;=\; i\,(k_0/n_\omega)E_\omega{}^* E_{2\omega}\, d_{eff}\exp(i\Delta kz)$$

and Harmonic:

$$\partial E_{2\omega}/\partial z \;=\; i\,(k_0/n_{2\omega})\, d_{eff}\, E_\omega{}^2\exp(-i\Delta kz)$$

where Δk is the mismatch of wave-vectors: $\Delta k = 2k_\omega - 2k_{2\omega} = (n_\omega - n_{2\omega})2\omega/c = (n_\omega - n_{2\omega})4\pi/\lambda_0$ and d_{eff} is the effective nonlinear coefficien and λ_0 is the free-space wavelength of the fundamental. Solving such equations demonstrates that power couples from the incident wave to its harmonic and then back again along the path length. Power couples into the harmonic for a length $(\Delta k/2)L_c = \pi$ and then begins to couple back again, so $L_c = \lambda_0/[2(n_0 - n_1)]$. Numbers for L_c can vary considerably, from less than ten microns to 100 microns or more. In quasi-phase matching periodic poling compensates every time the two waves reach a distance L_c. Solving the coupled mode equation yields the intensity at the second harmonic $I(2\omega)$ in terms of the intensity at the fundamental, $I(\omega)$ as a function of sample length l:

$$I(2\omega, l) = \frac{2\omega^2 d_{eff}^2 l^2}{n_{2\omega}n_\omega^2 c^3 \epsilon_0}\left(\frac{\sin\left(\Delta kl/2\right)}{\Delta kl/2}\right)^2 I^2(\omega)$$

where

$$I(2\omega, l) \ll I.$$

This is the familiar sinc^2 function; for a given crystal length l and intensity I, the maximum harmonic intensity falls off as the phase mismatch grows.

When phase-matched ($\Delta k = 0$), the harmonic intensity will grow quadratically with the fundamental intensity and quadratically with distance l, until the fundamental begins to deplete. It is then necessary to include depletion of the fundamental into the calculations. In the case of exact match, the solution is quite simple:

$$I(2\omega, l) = I(\omega, 0) \tanh^2 \left(\frac{E_0 \omega d_{\text{eff}} l}{n_\omega c} \right)$$

In the large conversion limit, $\tanh \to 1$ and the second harmonic intensity becomes 1; the fundamental will be completely converted to harmonic as expected by energy conservation (half the number of photons, but twice the frequency for each).

3. The P_3 term

The third order term produces a polarization proportional to third order in the electric field: $P_3 = \varepsilon_0(\chi_3 E^2)E$. This leads to a polarization oscillating at the third harmonic, and also to a term in which the factor ($\chi_3 E^2$) has no oscillation frequency; this term oscillates at the fundamental frequency of the incident light. Thus the third-order nonlinearity causes both third harmonic generation (THG) and also an intensity-dependent change in the refractive index, which becomes nonlinear, n(I). The third-order term does not require a material with a center of inversion symmetry; all materials have χ_3 terms.

3.1 Third and higher-order harmonics

Third-order nonlinearities arise from expanding the nonlinear susceptibility to third order in the electric field. The extension from second- to third- harmonic is relatively straightforward, at least in concept. Now there are three electric fields: E_0 oscillating at ω_0, E_2 oscillating at ω_2 and E_3 oscillating at ω_3. Depending on the medium, third harmonic can be considered as being generated by three fundamental photons (requiring phase-matching) or by a single incident photon interacting with the harmonic (a different kind of phase-matching). The detailed interactions are beyond the level of this chapter, but the full ramifications of third order nonlinearities were extensively discussed quite early (Armstrong, 1962; Maker, 1965; Hellwarth 1977). As soon as third harmonic is generated, the light intensity is usually high enough to generate even higher harmonics than the third and the Taylor expansion ceases to be a particularly valid way to look at these large nonlinearities. The main application of third harmonic is to reach ultraviolet (UV) wavelengths where there are few choices for lasers.

When the electric field strength of the light is high enough, optical nonlinearities can generate multiple orders with light having frequencies much greater than the original (up into the X-rays!). This is done using femtosecond pulses; proper phasing of the harmonics can lead to extremely short, extremely intense pulses, approaching attoseconds in length (10^{-18} s). This is the regime of extreme nonlinear optics (Song, 2010).

It is worth pointing out that phase-matching is not a particularly problem in gases, because their refractive indices are so small that dispersion is practically negligible. With high power short-pulse lasers, very high harmonics can be observed, out to the extreme UV (XUV, for λ < 100 nm) wavelength of 6.7 nm, which was achieved in a helium gas jet, starting with a KrF laser (Preston, 1996). The laser intensities were up to 4×10^{17} W/cm^2 in 380-fs pulses. This was the 37th harmonic.

High harmonic generation (HHG) requires high peak power, sub-picosecond lasers, which today are usually mode-locked Ti-doped sapphire (Ti:Al$_2$O$_3$) lasers. A moderate, commercially available system is capable of producing a sub-100 fs pulses with mJ pulse energies at pulse repetition frequencies of 1 kHz. This laser has revolutionized high field nonlinear optics because peak optical intensities of 10^{15} – 10^{18} W/cm^2 are routinely generated (Eden, 2004); specially designed gas jets enable maximum efficiency (Grant-Jacob, 2011). A later section will discuss special results in nonlinear media observed with pulsed lasers. Harmonic emission between 20 and 60 nm could be observed.

3.2 Nonlinear refractive index

When the nonlinear refractive index is nonlinear, the phase of the light changes. This does not require phase-matching, and so it occurs to some extent in all materials. In bulk materials, the nonlinearity in a plane wave cannot be directly observed because it is a pure phase-change. However, many geometries have proven to be important in producing nonlinear effects from the nonlinear refractive index, as will be described later. They include: 1) wave-mixing, when two plane waves intersect at angles in a nonlinear medium, which creates a phase-grating that can deflect (or reflect) light beams; 2) self-focusing or de-focusing of a beam of finite width; 3) formation of stable propagating beams called spatial solitons; 4) nonlinear waveguides; 5) nonlinear interfaces; and 6) in transient phenomena, where self-phase modulation can change monochromatic light into a frequency continuum.

The nonlinear refractive index, which depends quadratically on the light's electric field is usually written as n(I) = n$_o$ + n$_2$I, where I is the intensity of the light. Notice that in this nomenclature n$_2$ is related to χ$_3$. Because a nonlinear refractive index changes the phase experienced by the light, the conceptually simplest way to measure it is by interference with a reference beam. A more convenient experimental technique is wave-mixing (discussed below).

Alternatively, the nonlinear refractive index can be calculated from the Kramers-Kronig relation, which relates the nonlinear refractive index to the spectrum of nonlinear absorption, assuming the same level of excitation. The Kramers-Kronig relation states that when optical intensity I is present at frequency v, a change in refractive index Δn can be determined by measuring the change in absorption Δα at all possible frequencies v', under the same excitation conditions, and integrating by means of the following equation:

$$\Delta n(v) = \frac{c}{2\pi^2} \mathscr{P} \int_0^\infty \frac{\Delta\alpha(v')}{v'^2 - v^2} dv'$$

where $\mathscr{P}$ denotes the principal part of the integral. Because of the resonant denominator, the nonlinear refractive index due to nonlinear absorption may be quite large near a resonance. Indeed, in some geometries, such as an etalon, the nonlinear refractive index may have a bigger effect on the light beam than nonlinear absorption. Both the nonlinear refractive index and nonlinear absorption must be considered whenever an optical nonlinearity is near an absorption line.

3.2.1 Z-scan measurements of nonlinear refractive index

If the incident laser light is a Gaussian (or other shaped) beam, the nonlinear medium will induce a lateral spatially varying refractive index that will cause the beam to change its phase front. If the refractive index increases with intensity, an intense light beam will tend to "self-focus" as it travels through the medium (as discussed later). The nonlinear medium acts like a graded index lens. This lateral change in beam shape due to the nonlinear refractive index motivated a new technique called the *z-scan* that can measure both the magnitude and sign of the nonlinear refractive index (as well as the nonlinear absorption) (Sheik-Bahae, 1990).

In the z-scan measurement technique, a sample of the nonlinear material is moved through the focus of a laser beam, and an aperture is placed before the detector at some point in the expanding beam. The amount of light getting through this aperture is measured as a function of the sample position. If the nonlinearity is positive, the beam tends to self-focus, reducing the beam divergence and increasing the amount of light transmitted through the aperture. If the nonlinearity is negative, the amount of light will decrease. From the measured dependence of the detector signal on the sample position, it is possible to calculate the magnitude of the nonlinear index. The formula for the transmission is

$$T(z) = \frac{\int_{-\infty}^{\infty} P_r(\Delta\Phi_0(t))dt}{S\int_{-\infty}^{\infty} P_i(t)dt}$$

where P_T is the transmittance power through the aperture, which is a function of the phase distortion $\Delta\Phi_o$, and P_i is the incident power. Experimental data are fit to this equation. The fitting parameters are the nonlinear refractive index, together with nonlinear absorption.

3.2.2 Mechanism of nonlinear refractive index

The third-order nonlinearity produces a term that provides a nonlinear refractive index that is quadratic in the optical field (or linear in the optical intensity). This is sometimes called the *optical Kerr effect* (OKE) because the ordinary Kerr effect describes a refractive index change that depends quadratically on *applied* electric field.

The classical picture of a nonlinear refractive index is a nonlinear polarization, driven by the electric field. The polarization is proportional to the polarizability of the medium. In the simple physical picture, this is expressed as the oscillating charge separation introduced by the oscillating electric field: a simple harmonic oscillator. When the field is intense enough – approaching the value of the internal fields within the medium – the atomic cloud distorts and the harmonic oscillator becomes nonlinear. The nonlinearities can be calculated by a virtual mixing of the states of the atom or molecule. These classical and quantum mechanical pictures offer hints as to what materials will be the most nonlinear. However, for most applications, the nonlinear refractive index is determined by measurement. It is classically written as $n = n_o + n_2I$.

3.3 Cascading nonlinearity

It has already been pointed out that the effects at a given order of χ can be cascaded. For example, the χ_2 nonlinearity interacting with the fundamental can give rise to third harmonic. By the same token, one would expect cascading of χ_2 with the complex conjugate of the fundamental to give rise to an additional component at the fundamental frequency.

This shows up as a phase shift, which is equivalent to a nonlinear refractive index. Thus a nonlinear refractive index can be created by a cascading nonlinearity in a χ_2 material. Creating an effective χ_3 nonlinearity from χ_2 begins with phase-*mis*matched second-harmonic generation. Intensity-dependent up- and down- conversion both take place, providing an intensity-dependent phase change to the wave – thus mimicking a nonlinear refractive index. This can be understood as frequency-degenerate interactions between one wave (with itself) or two waves that induce self- or cross-phase modulation. (Stegeman, 1996; Lee, 2011)

4. Nonlinear absorption

Absorption processes can be strongly non-linear, particularly near atomic resonances. Assuming a nonlinear form of Beer's law, the intensity varies with distance z as $I = I_o\exp[a(I)z]$. The absorption coefficient α may become larger or smaller with increased intensity, depending on the physical process. Increasing absorption, even in transparent media, can come from the introduction of multi-photon absorption at high intensity levels. Reduced absorption comes from saturating the absorption line with high intensity light.

4.1 Multi-photon absorption

The absorption can be written heuristically as a function of intensity through $\alpha(I) = \alpha_o + \alpha_1 I + \alpha_2 I^2$ where α_o represents linear absorption, α_1 represents two-photon absorption, α_2 represents three-photon absorption, etc. The phenomenon of multi-photon absorption, where the absorption increases with intensity, is sometimes called *reverse saturable absorption*. Even though a material is transparent at low intensity, as the intensity grows, the absorption may increase. *Two-photon absorption*, which may occur in transparent materials, can be large when the sum of two photon energies comes close an absorption resonance (DeSalvo, 1996). This optical nonlinearity decreases the transmission of light as atoms or molecules absorb it while transitioning to higher levels. If the matrix elements allow it, these higher levels may return to the ground state by emitting fluorescence at twice the frequency of the input light. While this doesn't strictly follow the definition of nonlinear optics given in the first section, multi-photon fluorescence has become a very important tool for biological applications (Xu, 1996, Diaspro, 2006)).

Higher-order multi-photon resonant excitation processes may induce considerable absorption within the material. With enough laser intensity, it is possible to induce processes that mix quantum states of an atom or molecule at energies sufficiently high to cause ionization (Corkum, 1993; Guo, 2009). *Multi-photon ionization* is assumed to be the cause of the breakdown of air (or transparent materials) at the focal point of a high power light beam. Multi-photon absorption continues until ionization occurs and the material is damaged, often by the inclusion of microscopic bubbles. A commercial example that demonstrates this process is the transparent glass cubes (or other shapes) that can be purchased with 3D images written inside. This process is done with intersecting laser beams using multi-photon absorption that leads to ionization only where they overlap.

An important application of multi-photon absorption (or reverse saturable absorption) is *optical limiting*. This is defined as any process the limits the amount of light that can get through a material at high intensity. It has the potential application of protecting sensitive detectors and eyes from high power lasers (Tutt, 1993; Chi 2009; Kamina, 2009). With the advent of nanocrystals, research to find practical materials has exploded, yet to date it does not appear that any particular material stands out as viable.

4.2 Saturable absorption

When the frequency of incident light is near an absorption resonance of the material, the absorption may saturate as the intensity increases. Saturable absorption occurs when the incident intensity is high enough that the ground state population is depleted and the population of the upper and lower states equalizes. Although this is a microscopic process, it is often modeled heuristically by an absorption of the form $\alpha(I) = \alpha_o I_s/(I+I_s)$, where I_s is the *saturation intensity*. Saturable absorption has a number of practical applications, particularly for Q-switching and mode-locking lasers. Because the lowest loss occurs when the laser modes are locked together into pulses, introduction of a saturable absorber into a laser cavity enables it to passively mode-lock, an important way to generate ultra-short pulses. Saturable absorbers are also useful for nonlinear filtering outside laser resonators, which can clean up pulse shapes. Saturable absorption occurs at wavelength close to a resonance. Saturable absorption is particularly strong in semiconductor lasers at wavelengths just above the band edge.

5. Nonlinear scattering processes

In addition to transmission and absorption, light transmitted through materials can exhibit scattering. The linear elastic processes are Rayleigh and Mie scattering from density fluctuations in the medium. The non-elastic processes are Raman and Brillouin scattering. Rayleigh scattering comes from molecules and fine-scale density fluctuations (this is why the sky is blue); Mie scattering from fluctuations with a larger length-scale. Because the scattering particles do not move, light-induced density fluctuations appear constant in time (if the light is a continuous wave: cw) and appear as a light-induced refractive index change. These scattering processes are usually described by the χ_3 process explained earlier, so this section concentrates only on the non-elastic processes.

5.1 Stimulated Raman Scattering (SRS)

Raman and Brillouin scattering are inelastic events; the scattered light has a different frequency from the incident light. Spontaneous Raman Scattering usually down-shifts light to lower frequencies, because molecules in the medium begin to vibrate at frequency ω_r. If there are already oscillations in the medium, the light coming out will be up-shifted by the vibrational energy ω_r. Terminology has been developed that the down-shifted scattered components are called Stokes light, while the up-shifted components are called anti-Stokes light. (Stokes had already explained that in fluorescence, emitted light should always have a lower frequency than the incident light.) Raman spectroscopy has made important contributions to chemistry because it can identify molecular vibrations and measure their frequencies. These are spontaneous scattering events that can be understood only with quantum mechanics, and they have stimulated scattering analogs.

An analogy with lasers is very useful here; spontaneous emission in lasers is quantum-mechanical, while gain can be explained classically. Similarly, Raman and Brillouin scattering have their stimulated emission counterparts, offering gain. These processes can be understood classically when we assume that the scattered light is already present. We can then calculate the gain that the Stokes light experiences in the presence of the incident light. The existence of this gain, proportional to the incident light intensity, is what makes these nonlinear optical processes.

Assume the light field consists of incident and scattered components, E_0 and E_{-1} at frequencies ω_0 and ω_{-1}, respectively, where $\omega_{-1} = \omega_0 - \omega_r$. When ω_r is resonant with a molecular vibration, the frequency difference between these two waves can drive a molecular vibration. As with electronic states, this excited vibration can return to its non-vibrating state by spontaneous emission, or by stimulated emission.

When the incident light is intense and coherent, the molecular vibration is strongly driven at frequency ω_r as a result of interference between the incident and Stokes light (Garmire, 1963). The molecular bond length changes by an amount $\delta x \propto (E_0^* \bullet E_{-1})$, where the fields are considered vectors. The bond length oscillates at frequency ω_r, producing an oscillating electric-dipole moment that is proportional to the incident field through $\mu \propto E\delta x \propto E (E_0^* \bullet E_{-1})$. The dipole moment component at frequency $\omega_b - \omega_r$ is $\mu \propto (E_0^* \bullet E_{-1})E_0$, which drives power P_{-1} into the first Stokes through $P_{-1} = -(d\mu/dt) \bullet E^*$. This product means that the power delivered to Stokes light from the incident beam is proportional to the square of the Stokes field: $P_{-1} \propto |E_{-1}E_0|^2$. In other words, the gain is proportional to $|E_0|^2$, the incident optical power.

When the gain is weak, the Stokes light is spontaneous, emitted diffusely in angle and there is no *ab initio* phase relation between the Stokes and laser beam. The coherent molecular vibrations build up with a well-defined phase only when the laser beam intensity $\propto |E_0|^2$ becomes strong. Then the gain increases rapidly with incident light. For SRS to be practical for frequency conversion, a reasonable fraction of the incident light must be down-shifted by the SRS process. If equal powers in the incident and Stokes-shifted beams are assumed, in an interaction length L, the power requirement can be expressed as $P_0(L) = P_{-1}(L) = 16A_{eff}g_RL$, where g_R is the Raman gain and A_{eff} is the effective area of the beam (or fiber mode).

In Raman scattering, incident light induces molecular vibrations at natural resonance frequencies ω_r. These molecules have oscillating dipole moments given by the product of the molecular polarizability α_p and the electric field. When the light is coherent, the light-matter interaction transfers the phase coherence of the light to the molecules; the oscillating dipole moments induce a polarization density in the macroscopic medium, given by $P = N\alpha_pE$, where N is the density of oscillating molecules and E is the light's electric field at the molecule. An alternative picture describes this polarization density as a nonlinear susceptibility χ through $P = \varepsilon_0\chi E$. To first order, SRS can be expressed as a quadratic nonlinear susceptibility: $\chi = \chi_1 + \chi_3E^2$, where χ_3 is assumed to be highly resonant around the Raman resonance (and imaginary) (Shen, 1965).

As for anti-Stokes, if the electric field is assumed to additionally contain the frequency $\omega_0 + \omega_r$, then a coherent parametric interaction can take place. This interaction is out-of-phase relative to the incident field: $\mu \propto (E_0^* \bullet E_{-1})E_0^*$, which means that the anti-Stokes field drives the vibrating molecule back down to its ground state. This means that those periodic vibrations introduced into the medium by the emission of Stokes light are transferred back to the light wave in the form of coherent anti-Stokes emission – a classical resonant parametric process. For anti-Stokes to build up, the proper phase relation between the wave vectors must be maintained: $2k_0 = k_1 + k_{-1}$. When phase-matched, the power delivered to the anti-Stokes light is $P_{+1} \propto (E_0^*E_{-1})(E_0^*E_{+1})$, which is linearly proportional to the anti-Stokes field. The linear proportionality means that there is no threshold – the anti-Stokes field can grow solely from the interaction between the Stokes and incident field (as long as phase-matching the relationship is obeyed). This can be thought of as a *resonant four wave mixing process*.

Stokes radiation is usually emitted more strongly in the forward direction because the interaction length between the laser and Stokes is longest for the forward-directed components. The anti-Stokes is emitted in cones that obey the required phase relation with the forward-directed Stokes. When the Raman medium is inside the laser cavity, there is no anti-Stokes because there is no Stokes emission at the appropriate angles to feed into anti-Stokes.

The simple classical picture shows that higher order Stokes can be generated either by the first Stokes generating the second Stokes in exact analogy with the generation of first Stokes, or by a four-wave mixing process in which the molecular vibration due to E_0 mixes with E_{-1} to produce a dipole moment $\mu \propto (E_0^* \bullet E_{-1})E_{-1}'^*$, where the Stokes wave with field E_{-1}' does not have to travel in the same direction as the wave with field E_{-1}. The power generation at this second Stokes is $P_{-2} \propto (E_0^* \bullet E_{-1})(E_{-1}'^* \bullet E_{-2})$. This process has no threshold and requires the phase-matching wave vector relationship $k_0 - k_{-1} = k_{-1}' - k_{-2}$. This simple model also explains higher order anti-Stokes. Radiation at frequency $\omega_{+2} = \omega_0 + 2\omega_r$ is produced without threshold by modulation of the first anti-Stokes by the molecular vibration at frequency ω_r. The dipole moment $\mu \propto (E_0^* \bullet E_{-1})E_1^*$ drives second anti-Stokes power at $P_{+2} = (E_0^*E_{-1})(E_1^*E_2)$. This requires the phase-matching wave vector relationship $k_0 - k_{-1} = k_2 - k_1$.

The Raman gain is usually measured experimentally, and will be proportional to the spontaneous emission spectrum (as is true with lasers). Raman lasers occur when the laser-pumped Raman-active gain medium is placed between mirrors that reflect the first Stokes wavelength. Again, this is analogous to lasers, where the gain medium must be placed inside a cavity to achieve laser threshold. On the other hand, the Raman gain is very large, and can build up from noise to a substantial signal without any feedback. This strength makes SRS in Raman-active liquids very strong, and explains the deleterious SRS that builds up in fibers. Once the first Stokes has built up to a substantial signal, all the other wavelengths can build up from it by parametric processes.

Using solid-state terminology, molecular vibrations are optical phonons. With this point of view, it is reasonable to expect optically-excited acoustic phonons to have their stimulated counterpart. This is Stimulated Brillouin scattering, which will be discussed in the next section, but first we'll look at some SRS applications.

5.2 Applications of Stimulated Raman Scattering

A few exciting possibilities made possible with SRS are mentioned here. In most cases the value of SRS is that is produces coherent light at frequencies other than those available directly from known lasers. In fibers, where intense light travels for a long distance down a Raman-active medium (fused silica), SRS can build up and be detrimental by shifting the wavelength of the incident light. This is a problem in optical fibers for telecommunications and for flexible delivery of high-intensity light, although it is an advantage when the Raman light can be used as an amplifier. Spectroscopy is another field in which SRS has been particularly useful.

5.2.1 Raman lasers

Raman lasers come in many designs for obtaining new frequencies that are not produced by lasers themselves. The way to achieve Raman threshold with the lowest optical intensity is to put the Raman medium inside the laser cavity, utilizing the same cavity mirrors as the laser (with reflectivity at both the laser and Raman frequencies). For example, yellow light at 590 nm is a difficult laser color to produce, but can be achieved by frequency-doubling coherent light at 1180 nm with a KTP crystal. What source gives light at 1180 nm wavelength? A Raman laser from a barium tungstate (BaWO$_4$) Raman crystal placed inside a linear-cavity repetitively Q-switched diode-side-pumped Nd:YAG laser at 1064 nm. An average output of 3.14 W at 590 nm has been achieved (Li, 2007).

Hydrogen and methane under pressure offer the largest Raman shifts of all molecules and are used to obtain high-power pulsed lasers at new frequencies. For example, a frequency-doubled YAG laser Raman-shifted with methane provides output at 630 nm. Pressurized gases can be used in single pass, Raman resonator, oscillator-amplifier, and/or waveguide design; they don't have the self-focusing, Brillouin and anti-Stokes that solids have. They may be intra-cavity or extra-cavity.

Parametric Raman lasers can achieve efficient generation of both the second order Stokes and first anti-Stokes components emitting nearly diffraction-limited collimated beams. A first Raman laser is excited by some of the laser pump energy and produces a 1st Stokes component as a collimated beam. In the parametric Raman laser the Stokes beam interacts parametrically with the remaining collimated pump beam to produce high power (Grasiuk, 2004).

Raman lasers in silicon offer light generation within integrated circuit technology for intra-chip and chip-to-chip information transmittal. Raman lasers in silicon use a CW pump beam from a laser diode and have the first-Stokes laser mirrors integrated right into the silicon chip (Service, 2005), a technique necessitated because silicon itself cannot be made into a laser. Long-wavelength injection Raman lasers are composed of alloys of aluminum, gallium, indium, and arsenic. These Raman lasers are grown on a single chip; some layers convert electricity into an initial pump laser and other layers shift the light via first Stokes to longer wavelength, out to 9 μm (Troccoli, 2005).

5.2.2 SRS in fibers

The existence of SRS can be a problem in fiber optics, where over long distances it may broaden transmitted spectra. In other cases, SRS gain can be very useful, as an amplifier of weak signals, such as used in telecommunication systems. A Raman laser results when an SRS medium is placed in an optical cavity.

Stimulated Raman scattering places a serious limit to delivering power down a fiber. As shown earlier, the effective SRS threshold is P_{-1th} = 16 A_{eff} g_R L_{eff}, where now A_{eff} is the effective mode area of the fiber and L_{eff} is the length over which the SRS takes place. Larger mode area fibers enable more power to be sent down a given length without SRS becoming a problem, but they are not usually single mode, and aren't used in telecommunications. The Raman gain coefficient $g_R \sim 10^{-13}$ W/m for silica fibers (and polarized light).

While often detrimental, the SRS gain in fibers can be useful by providing a Raman amplifier, sometimes used in telecommunications. Distributed amplification is particularly appealing because it requires only a single pump source for all of the structure, reducing the network's cost and complexity. The down-side is that Raman amplification tends to require significant input powers. Distributed Raman amplification is applied in long-haul, broadband transmission systems that use wavelength division multiplexing (WDM) because amplifiers must provide a flat gain profile for all the signal channels. Raman amplifiers enable control over the Raman gain profile, and thereby reduce amplified spontaneous scattering noise. Adding a fiber Raman amplifier to a fiber Erbium amplifier offers a much wider gain bandwidth than either component alone (Islam, 2004; Headley, 2005).

Raman fiber lasers, with feedback to turn the amplifier into a laser, have potential applications in WDM systems, optical fiber sensors and spectroscopy. Continuously tunable channel spacing can be achieved with a hybrid of an erbium-doped fiber laser that gives high power conversion efficiency and a fiber Raman laser that has a large lasing bandwidth, both placed in an all-fiber ring cavity (Chen, 2007); stable multi-wavelength lasing has been observed over 24 wavelengths.

5.2.3 Stimulated Raman Spectroscopy

The nonlinear process of SRS provides a way to enhance the Raman signal for spectroscopic applications. The most common method is often called *CARS (Coherent anti-Stokes Raman Spectroscopy)*. Light at both the laser and first Stokes frequencies are incident on the Raman-active medium and the beating between these two laser beams sets up coherent molecular vibrations that parametrically generate the anti-Stokes frequency (Begley, 1974). This resonant four-wave mixing phenomenon obeys the usual requirement for phase-matching. The conversion efficiency from Stokes into anti-Stokes is proportional to the laser intensity squared, enabling as much as 10^5 times increase in the conversion efficiency. This method is particularly useful for investigating biological compounds where background fluorescence is a problem for conventional spontaneous Raman studies. CARS does, however, require a high power tunable laser (or an optical parametric amplifier). The spectroscopic techniques that use SRS are described in books and papers (Demtroder, 2008, McCamant, 2004).

Other stimulated Raman spectroscopy topics under study include transient behavior, Raman lasers, waveguides and fibers, and SRS at surfaces and in cavities. The ability of SRS to provide frequencies as needed has enabled coherent atom control (Hagley, 1999; Lukin, 2003) and manipulating quantum states (Bergmann, 1998; Scala, 2011).

5.3 Stimulated Brillouin Scattering (SBS)

In solid state physics, Raman-induced vibrations can be described as optical phonons. Interaction of light with acoustic phonons results in Brillouin scattering (the acoustic phonons are actually hypersonic, i.e. very high frequency acoustic waves). SBS can be understood, then, as a straight-forward extension of SRS to acoustic rather than optical phonons. SBS gain occurs in similarity with SRS. Stimulated Brillouin scattering occurs when two optical waves interact in materials through an acoustic wave generated by electrostriction. Electrostriction is the tendency of materials to become compressed in the presence of an electric field; electric fields change the density and therefore refractive index of electrostrictive materials. Spontaneous Brillouin scattering was known previously, just as spontaneous Raman scattering had been known before lasers. As with SRS, the spontaneous process is best explained quantum mechanically; an incident photon of energy hv_0 is scattered by an acoustic phonon of energy hv_a, and energy conservation requires that the scattered photon has an energy given by $hv_{-1} = hv - h_a v$. On the other hand *stimulated* Brillouin scattering can be explained classically as the beating of the electric fields of two optical waves that generate a hypersonic wave through electrostriction. This causes a moving index grating that scatters an incident optical wave and is the origin of the nonlinear coupling between the waves. The scattered wave frequency will down-shift or up-shift, depending on whether the acoustic phonon takes energy from the incident photon $h(v_0 - v_a)$ or gives its energy up to the incident photon $h(v_0 + v_a)$, respectively. There is a strong thermal excitation of acoustic phonons because the Brillouin shift is very small, so anti-Stokes up-shifted scattered light is comparable in intensity to Stokes down-shifted light. Typical Brillouin shifts are in the GHz regime. For example, the shift in optical fiber at wavelengths near 1.55 µm is ~ 9.6 GHz.

While SRS frequency shifts are given by fundamental characteristics of the scattering molecules and first Stokes has no phase-matching requirement, in SBS the acoustic frequency that determines the Stokes frequency shift is determined by a phase-matching requirement. This is because Raman-excited vibrations are localized on molecules, while Brillouin excitations are pressure waves that move with acoustic velocity v_a. The requirement that the hypersonic waves remain in phase with the interfering light waves provides the following vector relation: $k_{-1} = k_0 - k_a$. For the incident and scattered light to be in the same direction, the acoustic frequency would have to be zero. Because the magnitude of each vector is inversely proportional its velocity and the acoustic wave is much slower than the light waves (by a factor of 10^{-5}), the acoustic wave vector will be much larger than the optical wave vector, unless that acoustic frequency shift is very small. The largest the frequency shift occurs when the Stokes wave travels opposite the incident wave. In this case $\omega_a/\omega_0 = 2nv_a/c$, which is in the multi-GHz range.

5.3.1 SBS retro-reflection

The retro-reflection that occurs with SBS is perhaps its most important characteristic. The first experimental results on SBS excited in liquids (with a ruby laser) showed that the retro-reflected Stokes light returned to the (inhomogeneously broadened) laser, where this new frequency was amplified and reflected back to the liquid. Another SBS step then produced twice-shifted light (Garmire, 1964). It was only later that it was understood that *phase-conjugation* was taking place. A normal mirror changes the phase of the incident light upon reflection by π. Brillouin-reflected light, on the other hand, has a phase that is conjugate to that would be reflected off a usual mirror. This explains why SBS light could retrace its steps back into the laser, even after traveling through a lens.

To understand phase conjugation, consider the susceptibility that oscillates with the acoustic frequency by beating with a forward incident wave field $E_0 = \mathcal{E}_0 \exp[i(\omega t - kz)]$ and a backward-directed Brillouin wave field $E_{-1} = \mathcal{E}_{-1} \exp\{i[(\omega-\omega_r)t + k_{-1}z)]\}$. Considering complex notation, the term which gives the appropriate oscillation frequency for the susceptibility is $\chi = E_0 E_{-1}^*$, which oscillates at frequency ω_r, and has a rapid periodicity in z, due to the propagation of vector $k+k_{-1}$. The result is

$$\chi = \mathcal{E}_0 \mathcal{E}_{-1}^* \exp\{i[\omega_r t - (k_{-1} + k)z]\}.$$

The oscillating polarization that drives the backward wave E_{-1} is given by

$$P = \varepsilon_0\chi E_o{}^* = \varepsilon_0\, \mathcal{E}_0\varepsilon_0{}^*\mathcal{E}_{-1}{}^* \exp\{-i[(\omega-\omega_r)t + k_{-1}\, z]\}.$$

This a backward going traveling wave with a phase π different from the initial backward-going wave. This means the phase of this polarization density is conjugate from the initial Stokes field. This is the origin of the term "phase conjugation."

The "magic" of phase conjugation can be explained by looking at what the phase does for any wave leaving a point (x,y,z) on one side of the Brillouin phase conjugator. Assume the wave builds up a phase φ as it travels to the conjugator; when it reflects back, its phase in conjugated. That is, the phase φ becomes $-\varphi$. As that wave retro-reflects back to the point (x,y,z), it re-traces its steps and its phase returns to zero. This happens for every point, no matter what the phase distribution is between its and the conjugator. Thus any aberrations are completely cancelled out. If this were an ordinary mirror, the phase would *increase* from φ to 2φ upon a round-trip, rather than returning to zero. Thus the aberration *adds* for an ordinary mirror and *cancels* for a phase-conjugate mirror.

Nonlinear phase conjugation was first understood in the context of SBS and quickly was extended to other materials and processes, particularly photo-refractives, while SBS has remained a valuable way to reduce aberrations in high pulsed power applications, described later.

5.3.2 Performance limitations due to stimulated Brillouin scattering

In fiber telecommunications SBS puts a limit on the power that can be transmitted through fibers that is even more stringent than SRS. If the light in the fiber is too intense, SBS reflects light back where it came from, shifted down in frequency by the Brillouin acoustic vibration. This reduces the power that can be transmitted through single mode fibers for telecom applications (Shiraki, 1996). There is not a lot of flexibility in the design of telecommunications fibers because of requirements for low loss and low dispersion. The SBS gain G_B and threshold input (monochromatic) power P_{th} through a fiber of effective area A_{eff}, respectively, are given by:

$$G_B = (4\pi n_{eff}{}^8 p_{12}{}^2/\lambda^3\rho c v \Delta v_a)(P_o/A_{eff}) \text{ and } P_{th} = 21\, A_{eff}/g_B L_{eff}$$

where n_{deff} is the effective refractive index, p_{12} is the longitudinal elasto-optic coefficient, ρ is the density, c is the velocity of light (λ and v are the wavelength and frequency of the incident light, respectively), L_{eff} is the effective interaction length and Δv is the linewidth of the acoustic resonance (Kobyakov, 2005). The usual approach to reducing the effect of SBS is to use frequency-broadened pulses that smear the SBS gain over a range of wavelengths.

The gain is proportional to the intensity, (P_o/A_{eff}), as expected and proportional to the square of the elasto-optic coefficient. In narrow line fiber lasers and amplifiers, SBS remains the primary limitation on output power. Large mode area fibers decrease the optical intensity in the fiber core and raise the SBS threshold, but the maximum output power from narrow linewidth optical fiber amplifiers is still limited to approximately 100 Watts. Because SBS in an optical fiber occurs when the signal propagating in the core generates an acoustic wave that scatters light in the reverse direction, the SBS threshold can be raised by choosing a refractive index profile that minimizes the acousto-optic overlap while maintaining the desired optical properties (Kobyakov, 2005). Doping the core of the fiber with alumina (Al_2O_3) creates an optical waveguide but an acoustic anti-guide. Combining alumina and germania (GeO_2) doping in the fiber core can spatially separate the optical and acoustic fields, yielding over 500 Watts of power in a single-mode output, without the onset of SBS.

In laser-produced plasmas, SBS can be set up by thermal waves. In low-temperature, high-density high-Z plasmas this instability dominates and can produce significantly more SBS that expected (Short, 1992).

5.4 Applications of Stimulated Brillouin Scattering

Possible applications of SBS are too numerous to describe in detail. Indeed, several books on SBS have already been written (Damzen, 2003; Agrawal, 2008), describing the problems it causes in practical systems and how to overcome them, but also how SBS can be used to improve other systems. These positive applications broadly can be thought of as improving lasers and as improving sensing systems.

A phase conjugate mirror corrects wavefront aberrations, compensating for distortions of the laser beam created by inhomogeneities in the laser medium and/or its optical components. The SBS phase conjugate mirror is the simplest means to create phase-conjugation and is suitable for high power/energy laser systems. One problem needing correction by this means is thermal lensing caused by inefficient optical pumping of Nd:YAG lasers (Kovalev, 2005).

Phase conjugate mirrors are excellent for beam combining, although SBS requires the multiple lasers to be within the Brillouin gain linewidth. SBS phase conjugation is appropriate for combining beams from an amplifier array (Bowers, 1997). Research continues on how to combine the many beams needed for laser fusion (Kirkwood, 2011).

SBS can help clean up laser beams, because the backward-going Stokes has a much smoother beam profile than the incident laser beam. Thus it is sometimes practical, for example if using a multi-mode fiber, to use the output from a retro-reflected Stokes beam rather than the original laser beam (Steinhausser, 2007).

The SBS effect in a fiber ring sets up an acoustic wave that remains stable as the laser light beam travels around the ring, as long as the wavelength of the laser and the circumference of the ring are carefully matched. This enables the Brillouin laser to have an extremely narrow line, as narrow as 75 Hz (Geng, 2006).

Brillouin scattering depends on strain and temperature, making possible distributed sensing through Brillouin scattering in optical fibers. Brillouin enhanced sensing optical fibers can be imbedded in smart composites. The backward scattered Brillouin wave can travel over long fibers as an indicator of where the fiber is undergoing strain or other problems. One technique involves introducing pump and probe at both ends and doing an optical correlation. Both stimulated and spontaneous Brillouin scattering have been used for these applications (Horiguchi, 1995; Bao. 2011).

Laser pulse compression by SBS involves a tapered waveguide within a cell of pressurized methane gas. The Stokes pulses in the backward direction are compressed from incident nanosecond laser pulses (Hon, 1980). The taper ensures SBS starts at the far end of the cell and the transient dynamics between the incident, reflected and pressure wave all combine to reduce pulses to sub-nanosecond. With two cells in a generator-amplifier setup, up to 25 J in 15 ns pulses have been compressed to 600 ps (Dane, 1994); the two-stage process is much more stable than one alone (Erokhin, 2010).

SBS in photonic crystal fibers (PCF) can be dramatically altered by wavelength-scale periodic microstructuring, which alters both the optical and the acoustic properties. A PCF guides light through a lattice of hollow micro/nano channels running axially along its length. These fibers can be designed to either eliminate SBS or to increase it. The acoustic changes are particularly significant in fibers that contain filamentary voids. In one such example, the SBS threshold was increased five times when the Stokes frequency shift was in the 10-GHz range (Dainese, 2006).

6. Nonlocal optical/Photorefractive nonlinearities

Nonlocal phenomena occur when intense light entering the medium at one location in space changes the refractive index or absorption at nearby locations. Most typically this is due to diffusion of optically-induced excitation away from the initial point of excitation. A simple example is thermal nonlinearities, observed when thermal heating due to absorption of laser power spreads to adjacent areas, effecting the whole beam, not must the most intense parts of the beam. This was observed in increasing absorption in ZnSe waveguides, which exhibited optical bistability, which could not have occurred without the non-local nonlinearity (Kim, 1987). Most often, thermal nonlinearities cause *blooming* – in which a powerful beam spreads out ("blooms") because heating lowers the refractive index where the absorption of light is the strongest (in the center of the beam) – acting as a negative lens (Smith, 1977). The other main origin of non-local nonlinearities is the transport of optically induced charges in electro-optic media, which alter the refractive index. When this effect is detrimental, it is usually called "optical damage;" it ruins the spatial profile of the Gaussian beam. When these nonlinearities are wanted, to explore new phenomena, it is called *photorefractivity.*

6.1 Optical damage

High-power lasers can cause catastrophic optical damage by means of local bubble or crack formation, usually as a result of multi-photon ionization. The term *optical* damage is also used for non-catastrophic effects that medium-power lasers can introduce in electro-optic media. Such "damage" causes refractive index gradients that cause the beam to deform, interfering with its spatial profile or its waveguide properties (Mueller, 1984). This has been a particular problem in second harmonic generation, where the non-centrosymmetry requirement is also true for electro-optic coefficients. When it is useful, "optical damage" is typically called *photorefraction*; defects or atomic impurities in these crystals causes weak absorption of the light, liberating electrons which are free to move in the crystal, either by diffusion or drift. This separation of charges that occurs with the movement of electrons, creates internal electrical fields that, in turn, alter the refractive index of these electro-optic crystals. The altered refractive index affects the propagation of the light through the crystal.

The ability of light beams to create electric fields through charge separation is called the *photovoltaic effect*. The photovoltaic effect is most deleterious when the material is strongly insulating. Providing weak paths of conduction can remove the charge separation and the resulting photorefractivity. In LiNbO$_3$, addition of small amounts of MgO has been shown to reduce photo-refractivity by increasing conductivity.

6.2 Photorefractivity

In photorefractivity, the refractive index is locally modified by nearby spatial variations of the light intensity. Unusual new effects can be observed as a result of photo-excited carriers moving about in electro-optic crystals, due either to diffusion or to drift in local electric fields. The strongest effects are observed when coherent waves interfere to form a spatially varying pattern of illumination. As a result of photo-excited charge migration, a space charge is introduced that results in an electric field that changes the refractive index via the electro-optic effect (Cronin-Golomb, 1984). Two light beams interfering in a photorefractive medium generate photo-carriers in the spatially periodic bright regions. These carriers move to the spatially periodic dark regions where they are trapped. These trapped charges introduce a periodic electric field that creates a periodic refractive index distribution if the material is electro-optic. This refractive index grating is spatially shifted from the incident interference pattern and can diffract light into new directions. This can occur at quite low optical power levels, although it may take some time for substantial charge distributions to build up. Applications include two-beam coupling, dynamic holography, phase conjugation and spatial solution formation (Gunter, 1982).

Photo-refractivity was first discovered in lithium niobate, where it was shown that holograms could be written in real-time in the crystal, which offered promise for image processing. New crystals were investigated for photorefractivity and barium titanate was found to be have a large nonlinearity, resulting in interesting nonlinear effects, particularly related to phase conjugation. Photorefractive crystals have the advantage of high sensitivity, but tend to be very slow. In barium titanate the effects take seconds to build up; also it is sensitive only in the blue (Chang, 1985; Feinberg, 1980).

Photorefractivity has been extended to semi-insulating III-V and II-VI semiconductors, where the effect is not as large, but can be a thousand times faster, and the light source can be in the infrared – even at a wavelength of 1.55 microns in CdTe (Partovi, 1990). The largest effects require kilovolts to be applied to the crystal, but kilohertz response can be achieved. If the wavelength is near the band edge of the semiconductor, the local resonant nonlinear refractive index can add to the non-local electro-optic refractive index change to enhance the photorefractivity (Partovi, 1991). Wave-mixing can be observed with mW of incident power (Nolte, 1999).

More recently photorefractive polymers have proved effective, with the potential of low cost real-time holography (Ostroverkhova, 2004).

6.3 Photo-refractive materials

In the photorefractive effect, the local index of refraction is modified by spatial variations of the light intensity. It is typically most useful when coherent beams interfere with each other to form a spatially varying pattern of illumination. As a result, charge carriers are produced in the material, which migrate owing to drift or diffusion and space charge separation effects. The resulting electric field that is produced induces a refractive index change via the electro-optic effect. Materials must have a large electro-optic and photo-induced charge carriers.

6.3.1 Photorefractive crystals

The least expensive and commonly used photorefractive crystal, which has been around for a long time, is iron-doped lithium niobate (Fe:LiNbO$_3$), which can also be doped with titanium or cerium. It has large electro-optical (EO) coefficients, high photorefractive sensitivity and diffraction efficiency, but the photorefractive effect in this material is very slow, because charge transport is slow. Photo-refractive phase-conjugation was first observed in Barium Titanate (BaTiO$_3$). Because it is difficult to grow and delicate to handle, barium titanate has largely been supplanted by other, more practical, crystals. Lithium niobate, for example, is more suitable for volume fabrication and practical devices.

Strontium-Barium Niobate (Sr$_x$Ba$_{(1-x)}$Nb$_2$O$_6$) SBN is an excellent optical and photorefractive material which can be nominally pure or doped with Ce, Cr, Co or Fe. SBN has a large electro-optic coefficient delivering a fast response time and high two- wavelength mixing gain. Its figure of merit for photorefractive applications is much larger than lithium niobate, opening the way to much smaller devices. No applied field is required to enhance two-beam coupling.

Sillenite single-crystal bismuth silicon oxide Bi$_{12}$SiO$_{20}$ (BSO) and bismuth germanium oxide Bi$_{12}$GeO$_{20}$ (BGO) show a unique combination of different physical properties. These are the fastest photorefractive crystals to date. The coupling gain can be enhanced by applying an external electric field. The cubic crystalline structure is enables polarization manipulation in 2 and 4 wave mixing configurations. The crystals are very efficient photoconductors with low dark conductivity that allows a build-up of large photo-induced space-charges. These materials make possible a wide range of optical devices and systems for spatial light modulators, dynamic real-time hologram recording devices, phase conjugation wave-mixing, optical correlators, and optical laser systems for adaptive correction of ultrashort light pulses. These materials can be produced as thin-film structures for optical waveguide and integrated optical devices, as well as in bulk.

6.3.2 Photorefractive semiconductors

Several semi-insulating compound semiconductors have been demonstrated to be photorefractive. They include undoped and chromium-doped gallium arsenide (GaAs, GaAs:Cr), iron-doped and titanium-doped indium phosphide (Fe:InP, Ti:InP), undoped gallium phosphide (GaP) and vanadium- and titanium-doped cadmium telluride (CdTe:V, CdTe:Ti). These photorefractive semiconductors provide several attractive features for information-processing applications and could lead to a new generation of integrated optical information processors.

The semiconductor material must have adequate densities of localized energy levels to act as donors and acceptors for supplying and receiving the transferred charges, respectively. Furthermore, the photorefractive material has to be insulating or semi-insulating in order to avoid Coulomb screening around the charged centers. Typical resistivity of semi-insulating GaAs is higher than InP, due to its higher bandgap; semi-insulating CdTe has even higher resistivity. A commonly used figure of merit is n^{3}r/ε, where r is the electro-optic coefficient. The figures of merit for GaAs, GaP, InP, and CdTe are 3.3, 3.7, 4.1, and 16, respectively. Other crystals, BSO, SBN, BaTiO$_3$, LiNbO$_3$ and KNbO$_3$, have figures of merit of 1.8, 4.8, 4.9, 11, and 14, respectively. CdTe has the highest figure of merit among all the materials listed here. In many of the semiconductors, photo-refractivity can be larger if an external field is applied. The big advantage of semiconductors is that their response time is high; they turn off much more rapidly than the crystals. (The turn-on time depends on incident intensity.)

6.3.3 Photorefractive organic materials

The main classes of photorefractive (PR) organic materials include polymer composites, small molecular weight glasses, fully-functionalized polymers, polymer-dispersed liquid crystals, liquid crystals and hybrid organic-inorganic composites. The best performing photorefractive organic materials exhibit two-beam coupling gain coefficients Γ = 200-400 cm^{-1}, giving nearly 100% diffraction efficiencies in rather thin films. Their grating formation times are on the order of several milliseconds (Eralp, 2006). The advantage of polymer composites is the ability to tune their photorefractive properties by varying the concentration and type of constituents. However, because many components are combined in the composites, phase separation and crystallization can reduce the shelf life of devices. Also, while adding a plasticizer enhances chromophore orientation, it also increases the inert volume, which reduces overall photorefractivity (Grazulevicius, 2003; Marder, 1997). Organic glasses resolve these issues, but reduce the flexibility to tune the material's properties (Zhang, 2011). Both the shelf life of photorefractive polymers and the quality of starting materials available remain problems that have until now kept photorefractive organic materials as research materials.

6.4 Photo-refractive applications

Proposed applications include read-time holography, optical image processing, high density optical data storage, optical computing, communications, image processing, neural networks, associative memories, phase conjugation, laser resonators, and many others. Image processing applications include image correlation, image amplification, and dynamic novelty filtering. Data can be stored in photorefractive materials in the form of 3D phase holograms that have very high density and fast parallel optical access. Phase-conjugation has been used to correct image distortions suffered by optical beams in inhomogeneous or turbulent media. Photorefractive crystals, semi-insulating semiconductors, and polymer films have all been used to demonstrate proposed applications, but there remain no widespread commercial applications, due to the high cost of quality crystals and performance limitations, particularly the slow hologram formation speed in most materials.

Some photorefractive applications are on the horizon. Perhaps one of the most exciting is real-time holography that can display people, objects or scenes in three dimensions. The holograms can be seen with the unassisted eye and are similar to how humans see their actual environment. The concept of 3D telepresence, a real-time dynamic hologram depicting a scene occurring somewhere else, is surey an application with promise. A holographic stereographic technique is used, along with a photorefractive polymer material as the recording medium. The holographic display refreshes images every two seconds. A 50 Hz nanosecond pulsed laser writes holographic pixels. Multicoloured holographic 3D images are produced by using angular multiplexing, and the full parallax display employs spatial multiplexing (Blanche, 2010). Such applications will certainly increase in the future as materials become better.

6.5 Charge transport nonlinearities

A range of other nonlinear phenomena depend on the spatial transport of optically induced charge carriers within internal electric fields, particularly in semiconductor devices (Garmire, 1989). Schottky barriers and *pn* junctions provide internal fields that can be depleted by motion of photo-carriers, offering exquisitely sensitive nonlinearities through the electro-optic effect and through band-filling (Dohler, 1986; Jokerst, 1988). The modulation-doped n-i-p-i and hetero-n-i-p-i structures are examples, with the nonlinear refractive index due to resonant phenomena that can respond in milliseconds to microwatts of optical power (Kost, 1988).

7. Wave-mixing in nonlinear materials

Wave-mixing geometries involve two or more plane-waves, incident at an angle in a bulk nonlinear medium. The nonlinearity can be absorptive or it can rely on a nonlinear refractive index (or both). When the medium is photo-refractive, with mobile optically-excited charge carriers, the phenomena are particularly interesting.

7.1 Two-beam coupling

When two coherent beams interfere in a nonlinear medium, their interference introduces a grating in absorption or refractive index inside the material. These beams do not interact if the optical nonlinearity is local; they merely pass through each other. If photo-induced charges move within the material, however, a grating is set up that moves laterally with respect to the incident beams. This grating can diffract one beam into the other. The direction of energy transfer is determined by the sign of the mobile charge carriers and the electro-optic response. The energy transfer direction can be reversed by changing the polarity of the electric field (Partovi, 1987; Kim, 2011).

Two-beam coupling makes possible the amplification of a weak beam by means of coupling from a strong beam. Analysis of the coupled mode equations shows that the ratio of the weak beam to the strong beam increases exponentially with two-beam-coupling gain-length product ΓL:

$$I_{wo}/I_{so} = (I_{wi}/I_{si})\exp(-\Gamma L) \text{ where } \Gamma = (4\pi/\lambda)(\Delta n/m) \sin\Phi.$$

where I_{wo} and I_{so} are the weak and strong output beams, respectively, and I_{wi} and I_{si} are the weak and strong incident beams, respectively. Δn is the peak nonlinear refractive index introduced by the interference, m is the diffraction order, and Φ is the lateral phase angle between the periodic spatial intensity profile and the refractive index grating (that was moved by photo-charge transport). The gain can be used as a polarization-converter if the unwanted polarization is used to drive the gain (Heebner, 2000).

Beam coupling has been observed in the conventional photo-refractive crystals such as barium titanate and $Fe:LiNbO_3$. Some possible applications are optical limiting, produced by some fraction of the pump beam being reflected back onto itself, thereby robbing the incident beam if it becomes too powerful.

7.2 Three-wave mixing

Parametric processes involve three waves interacting. An example is parametric down-conversion, which is at the heart of the optical parametric oscillator (OPO) that will be discussed later. In a χ_2 medium, each incident pump photon breaks up into two less-energetic photons (the signal and the idler) such that the sum of their energies equals that of the pump photon. The sum of the signal and idler wave-vectors must also equal that of the pump ("phase-matching," as required in second-harmonic generation).

Parametric down-conversion can be regarded as the inverse process of sum-frequency generation, in which two beams at different frequencies create a beam that has a frequency equal to the sum of their frequencies. Thus sum- and difference- frequency processes are also three-wave mixing processes. In difference-frequency generation, it can be considered that both the pump beam and an intense idler beam mix create the signal beam. In parametric generation, the idler signal builds up from noise and feeds the signal beam with a gain per unit length. This process is enhanced by providing a cavity for the signal beam, in which case it has a threshold and becomes an optical parametric oscillator (OPO). These processes are particularly important to reach wavelengths in the mid-infrared.

The mixing of an anti-Stokes Raman wave with the laser beam and a Stokes wave is another three-wave mixing process. These processes do not have thresholds (unless they are placed in cavities). Three-wave mixing can even be used for generating holograms (Bondani, 2002). Three-wave mixing can be effectively applied to wavelength conversion, all-optical gating, all-optical switching, optical parametric amplification and oscillation, where it can increase the wavelength range over which these applications can operate (Liu, 2002).

7.3 Four-wave mixing

Four-wave mixing is a nonlinear effect arising from a third-order optical nonlinearity, χ_3. The four-wave mixing (FWM) geometry is similar to two-beam coupling: two incident (or "pump") beams (or "waves") "write" an optically-induced grating by means of nonlinear refractive index, non-linear absorption, or both. In FWM there is also a probe, or reading, wave that is partially diffracted from the optically induced grating to form a fourth wave, called the "signal beam." In the degenerate FWM geometry (DFWM), all four beams have the same wavelength. FWM provides background-free detection of very weak diffraction signals, since the signal wave appears at a different angle from the rest of the light.

The diffracted beam intensity (signal) is typically measured as a function of time, applied electric field, writing beam intensities, etc and the diffraction efficiency is determined. Typically the probing beam should not disturb the grating, which is achieved by making the probe beam much weaker than the pump beams and/or by having the probe beam polarized orthogonal to the writing beams. In the approximation of thick (volume) grating, the diffraction efficiency (signal intensity divided by incident intensity) in a sample of length L is given by:

$$\eta = \exp(-\alpha L)\sin^2(\pi \Delta n L / \lambda)$$

where αL is any residual (linear) absorption and Δn is the maximum amplitude of the refractive index grating, where it is assumed that the incident and diffracted fields are parallel. Note that when the refractive index modulation Δn is small, the diffraction efficiency is proportional to $(\Delta n L)^2$.

If the two waves that interfere are at the same frequency, the grating is stationary. In the thick grating limit, the Bragg condition must be satisfied, so in DFWM the probe beam must enter at the same angle as one of the pump beams. If DFWM takes place in a thin film, then the Raman-Nath condition holds and the probe beam can be at any angle. If the probe beam is a different frequency from the pumps and the grating is thick, the Bragg angle must be chosen: $\sin\theta_{probe} = n\lambda_{probe}/2d$, where d is the grating spacing caused by interference of the two pump beams.

When the two pump beams are at different frequencies, then the refractive index grating will move back and forth laterally. The probe-wave reflecting off this grating will be frequency-shifted by the frequency of the moving grating, just as in an acousto-optic modulator.

Many FWM experiments are possible: FWM measures lifetimes of gratings, spatial motions, surface effects, etc. (Abeeluck, 2002). Pulsed pump and probe waves, with a time delay between them, is a particularly valuable way to measure transient phenomena, both excited state lifetimes and dephasing (Yang, 1994). Many optical nonlinearities can be explained under the general concept of FWM, such as some third harmonic processes, SRS, SBS, parametric amplification, photo-refractive effect, and self-phase modulation (discussed later). The concept is useful to understanding a variety of spectroscopic tools, also discussed later. The most significant of these is CARS (coherent anti-Stokes Raman spectroscopy) where two input waves generate a detected signal with slightly higher optical frequency due to internal molecular vibrations. With a variable time delay between the input beams, it is also possible to measure excited-state lifetimes and dephasing rates (Becerra, 2010).

In fibers, FWM can be both a blessing and a curse. Non-degenerate FWM occurs in a fiber when two or more different frequencies propagate together, due to the fact that doped silica has a χ_3. With two input frequencies, a refractive index modulation at the difference frequency occurs, which creates two additional frequency components as sidebands on the initial waves. If there is already light at these sideband frequencies, it can be amplified, i.e., it experiences parametric amplification. In this way FWM may be a detriment to optical communications. As with SRS and SBS, FWM can be useful or harmful, depending on the application. Four-wave mixing in fibers is related to self-phase modulation and cross-phase modulation, transient effects that will be discussed later; this leads to spectral broadening that is particularly deleterious to WDM (wavelength division multiplexing), where it can cause cross-talk between different wavelength channels, and/or an imbalance of channel powers. One way to suppress this is to avoid equidistant channel spacing.

7.4 Phase conjugation and its applications

When two waves are counterpropagating ($k_1 = -k_2$) in a nonlinear medium, their interference sets up a grating with a periodicity of a half-wavelength. When a third wave at the same frequency is incident on this grating, the fourth wave that constitutes its reflection ($k_4 = -k_3$) will be the phase conjugate of the third wave. As discussed in SBS, where phase-conjugation is also seen, phase conjugation results in a retroreflection that overcomes aberrations. Photorefractive materials enable phase conjugation at relatively low optical power levels (Yariv, 1978).

The phase-conjugating grating exactly reverses the phase of any third wave. Thus, if a beam has gone through an aberration and forms a distorted wave, each portion of this wave will have its phase reversed upon reflection, so that its path will exactly reverse, re-creating the original beam's spatial profile after the waves pass through the aberrator. This phenomenon occurs in SBS and also in photorefractive materials.

Practical applications considered for phase conjugation include correcting wavefront aberrations in a laser beam (Bach, 2010). In optical beam clean-up, the signal beam that contains information about the object is combined with the reference beam in a photorefractive material, and a volume hologram of the object is recorded. If the signal beam went through an aberrator (which would correspond to the situation when the object has to be imaged through a medium with turbulence, refractive index inhomogeneities, etc.), then the image would be heavily distorted. If, however, a reading beam counterpropagating to the reference beam is introduced, it generates the phase-conjugated replica of the signal beam, which retraces its path through the aberrator, creating a cleaned-up image of the object. Impressive demonstrations have been provided, but practical systems for such applications have not yet been developed.

In the future it may be possible to use phase conjugation to transmit undistorted images through optical fibers (or the atmosphere), to provide lensless imaging down to submicrometer-size resolution to improve optical tracking of objects, phase locking of lasers (although multiple lasers must be close to the same frequency and combining laser beams hasn't yet been shown practical), refreshing of holograms for long-term optical storage, optical interferometry, and image processing. Phase conjugation has demonstrated a number of remarkable phenomena, particularly related to image processing (novelty filtering; edge filtering etc.), but to date only hero demonstrations have been reported. A number of practical issues still must be solved before phase conjugation is likely to be useful. The most likely application lies in telecommunications where it has been shown that nonlinear phase noise is effectively compensated in a midlink optical phase conjugation configuration (Jansen, 2006).

8. Transient nonlinear optics

Some phenomena occur only in the transient regime. These have become particularly important because ultra-short laser pulses (as short as femtoseconds) can provide changes in optical fields faster than any characteristic times in the system. Many of these effects, such as self-induced transparency (Fleischhauer, 2005) and photon echoes (Zewail, 1980; McAuslan, 2011), require quantum mechanical coherence of atomic states and will not be discussed here.

8.1 Self-phase modulation

Self-phase modulation and cross-phase modulation are important transient phenomena that must be included. These phenomena rely on the fact that a pulse traveling through a nonlinear medium sees a time-dependent refractive index, due to the fact that the intensity changes over the time of the pulse. And a time-dependent refractive index introduces a time-variable phase shift that broadens the frequency spectrum of the pulse (Genty, 2007). Because the nonlinear refractive index depends on intensity, when the intensity depends on time through $I(t)$, so does the refractive index, $n(t)$. To first order, its time dependence can be written as $n(t) = n_0 + t(\partial n/\partial t) = n_0 + tn_2 \partial I/\partial t$. A linear variation in time can be considered a frequency shift, so the nonlinear refractive index causing the light to modulate its own phase means that the light undergoes a frequency shift given by $\Delta v = n_2(\partial I/\partial t)(z/\lambda_0)$. At the beginning of a light pulse, the intensity is small, it rises to a maximum and then returns to zero. Thus the phase varies during the duration of the pulse and generates a continuum of frequencies. Pure SPM broadens the frequency spectrum of the pulse symmetrically, introducing a pure phase shift; it does not change the envelope of the pulse in the time domain.

In any real medium, however, dispersion will also act on the pulse. In regions of normal dispersion, the "redder" portions of the pulse have a higher velocity than the "blue" portions, and thus the weaker part of the pulse moves faster than its stronger parts, broadening the pulse in time. In regions of anomalous dispersion, the opposite is true, and the pulse is compressed temporally and becomes shorter. Using femtosecond lasers in specially designed fibers, self-phase modulation can be so large as to produce a white-light continuum which has proven to be very useful for spectroscopy Ranka, 2000).

If the pulse is strong enough, the spectral broadening process of SPM can balance the temporal compression due to anomalous dispersion and reach an equilibrium state, called an optical temporal soliton, discussed later.

Thus we see that self-phase modulation can introduce spectral broadening, extending all the way to a supercontinuum in fiber, or it can compress the pulse in time. When designed properly, self-phase-modulation can also narrow the spectrum (which widens the pulse in time).

8.2 Cross-phase modulation (XPM)

Cross-phase modulation is a direct analog of self-phase modulation in which light at one wavelength can change the phase of light at another wavelength of light through the optical Kerr effect χ_3.

Cross-phase modulation means that different laser pulses within a medium can interact. It is possible to determine the optical intensity of one pulse by monitoring a phase change of the other one. Because no photons in the first beam are absorbed, such a measurement is called quantum nondemolition (QND). XPM can be used to synchronize two mode-locked lasers that co-exist in the same gain medium, as long as the pulses overlap and experience cross-phase modulation. In optical fiber communications, XPM can cause channel cross-talk, particularly in DWDM (dense wavelength division multiplexing) systems. Cross-phase modulation has been considered for wavelength conversion in optical communications, usually based on changes in the refractive index via the carrier density in a semiconductor optical amplifier.

Cross-phase modulation can be used as a technique for adding information to a light stream by modifying the phase of a coherent optical beam with another beam through interactions in an appropriate non-linear medium.

In DWDM applications with intensity modulation and direct detection of several wavelengths transmitted simultaneously, first the signal is phase-modulated by the co-propagating second signal. In a second step dispersion leads to a transformation of the phase modulation into a power variation. The presence of dispersion in this case results in a walk-off of pulses between the channels and reduces the XPM-effect.

8.3 Temporal solitons

A soliton is a wave with a unique shape that travels undisturbed without changing. Solitons require a balance between nonlinearity and dispersion. Optical solitons are most often thought of in the time domain, especially in fibers, where pulse compression due to the optical nonlinearity can overcome the tendency of dispersion to spread pulses out as they travel down a fiber. This happens only for discrete values of the pulse energy and, for positive nonlinearities, the dispersion must be anomalous (negative) (Haus, 1996). Solitons are also called solitary waves and were first discovered in water waves traveling down a canal. Solitons are unique in that they can interact with other solitons and emerge from the collision unchanged, except for a phase shift.

Thus and optical temporal soliton is a pulse of light traveling (usually down a fiber) at its group velocity, while maintaining its same shape. Solitons occur because of an exact balance between dispersion (that tends to spread out the pulse in time) and self-phase-modulation (that tends to widen the spectrum and narrow the pulse). Solitons can be found by solving the non-linear wave equation in the presence of dispersion. The unique time-dependence of the electric field amplitude of the pulse has the form: $E(L,t) = E_o \, sech(t/T_o)$.

This pulse has a unique amplitude; its peak intensity is linearly proportional to the group velocity dispersion (GVD), $|\partial^2\omega/\partial k^2|$, and inversely proportional to the pulsewidth squared, as shown: $I = n_o \, |\partial^2\omega/\partial k^2| \, /(\omega_o n_2 v_g^2 T_o^2)$.

Solitons can be explained by considering that the chirp produced by SPM, with high frequencies in back and low frequencies in front, is offset by dispersion, which slows the low frequencies in front of the pulse and speeds the high frequencies in the back of the pulse. The resulting pulse does not change its shape as it travels down the fiber; the dispersion is kept in balance by the nonlinearity and vice versa. If an input pulse does not have the exact soliton shape, a clean soliton will eventually emerge after the undesirable portions of the excitation spread out in time.

Higher order solitary waves exist. An $N = 2$ soliton starts out as a simple pulse, but as it travels it sharpens in time while developing side-peaks. It fully recovers after a certain period, only to restart the process. The $N = 2$ soliton requires approximately twice the intensity of the $N = 1$ soliton. Solitons higher than $N = 2$ always have multiple peaks in time. These peaks have interesting behaviors, such as passing through each other without interfering. The solutions to the nonlinear wave equation rapidly become very complex and will not be further considered here.

9. Beam-related non-linear effects

To this point we have generally assumed plane waves, either a single wave or interfering waves. (This assumption was violated, however, when describing the Z-scan method for evaluating optical nonlinearities.) A number of interesting phenomena occur when a Gaussian beam, or any other beam of finite width, travels through a nonlinear medium. These effects include self-focusing and optical solitons.

9.1 Self-focusing

When an intense beam is focused into a material with a nonlinear refractive index, the phase velocity decreases with increasing intensity near the center of the beam. This means equiphase surfaces are compressed near the axis where the beam is more intense. Since rays are normal to the equiphase surfaces, they will tend toward the region of highest intensity, coming to a focus if they can overcome diffraction. This tendency to self-focus is offset by the tendency to diffract. Thus the critical power must be high enough that self-focusing wins.

Self-focusing occurs if the radiation power is greater than a critical power value $P_{cr} = \alpha\lambda^2/(4\pi n_o n_2)$, where λ is the radiation wavelength in vacuum and α is a constant that depends on the initial spatial distribution of the beam and it is approximately 2 for Gaussian-shaped profiles. Of course this critical power depends on the nonlinear coefficient n_2. For air, $n_2 \approx 4 \times 10^{-23}$ m^2/W for $\lambda = 800$ nm, and the critical power is $P_{cr} \approx 2.4$ GW, corresponding to an energy of about 0.3 mJ for a pulse duration of 100 fs. For fused silica, $n_2 \approx 2.4 \times 10^{-20}$ m^2/W, and the critical power is $P_{cr} \approx 1.6$ MW.

When a beam enters a nonlinear medium, a simple model of a uniform beam of radius a entering a nonlinear medium were $\Delta n = n_2 I$, predicts that it will take a distance z_f before self-focusing occurs, where $z_f{}^2 = (a^2/4)n_o/\Delta n$ (Garmire, 1966). When intense light self-focuses, the intensity becomes large enough that all sorts of additional nonlinearities become large, particularly SRS and SBS. If the beam has hot-spots, the self-focusing action can cause it to break up into filaments (Brewer, 1968). Local areas can become bright enough that they can damage the material. Self-focusing is a real problem that must be overcome in high power nonlinear systems.

9.2 Spatial solitons

The spatial soliton concept preceded self-focusing but is much harder to create in the laboratory. When diffraction and self-focusing are balanced, the nonlinear medium can cause the optical beam to trap itself. As a simple estimate, consider the diffraction of a circular optical beam with a uniform intensity profile in a nonlinear material whose refractive index varies with intensity as $n = n_o + n_2 I$. In a linear medium a beam of diameter D is expected to diffract with angle $\theta_D = 1.22\lambda/n_o D$. For a sufficiently intense beam, the nonlinearity can cause a large enough dielectric discontinuity at the edge of the beam that the critical angle for total internal reflection θ_C is greater than θ_D and the beam cannot diffract. For larger diameter beams, the critical angle θ_C becomes smaller, but so does the diffraction angle θ_D. For smaller diameter beams, θ_C is larger, but so is θ_D. This means there is a particular value of the power, independent of the diameter of the beam, at which we expect to see the beam self-trap. This value is $P_{cr} = 1.22^2 \pi\lambda^2/32 n_2 n^2$. (Chiao, 1963, Wright, 1995) The shape must be calculated numerically, but it is clear that the beam area and peak intensity are inversely related, in order to hold the power constant.

For a slab beam there is an analytic solution for the 1D soliton, in which the ideal beam intensity profile is $\text{sech}^2\Gamma y$, for which there is a critical power, given by $P_{cr} = 2\pi/n_2 n k_o{}^2$, assuming the nonlinear refractive index is expressed as $n_2 I$. The size of the beam is determined by Γ, which is approximately the inverse beam-width, and is given by $\Gamma = \frac{1}{2}\varepsilon_2{}^{1/2}k_o E_t(0)$, where $E_t(0)$ is the peak value of the optical field.

In a pure χ_3 nonlinearity (optical Kerr effect), the 2D cylindrical beam soliton turns out to be unstable at the critical power, resulting in filamentation and multiple self-focusing. Nonetheless, stable solitons can be created in media with a saturating nonlinearity, or in media with a χ_5 term. In Kerr media, 1D solitons (with a slab beam) are stable. Stable 1D solitons lasting as long as 5 cm have been reported in carbon disulfide (Barthelemy, 1985). Stable self-trapping can also be observed in the plane of a nonlinear waveguide (Stegeman, 1986). Waveguide confinement out-of-plane means the optical field follows the 2-D solution of the nonlinear wave equation, enabling the beam to travel stably, without diffraction or focusing, in a special in-plane spatial distribution. At higher power, the characteristics of higher-order spatial solitons can be seen (Maneuf, 1988). An amplified mode-locked dye laser with 75 fs pulse-length can to trap itself in a spatial soliton in a glass waveguide 5 mm long (Aitchison, 1990); the small nonlinear coefficient of glass requires high peak power to trap the beam. The guided beam can retain its original 15 µm width at a peak power of $P \sim 400$ kW. While an impressive result, these powers are much too high to be practical.

Spatial optical solitons are possible in a χ_2 medium using the cascading nonlinearity discussed above (Torruellas, 1995). With phase matching, both the fundamental and second harmonic can be mutually trapped.

9.2.1 Spatial solitons in photorefractive media

Spatial solitons can occur in photorefractive media, where the critical power can be on the order of 10 µW (intensities of about 200 mW/cm^2) (Duree, 1993). Photorefractive crystals like SBN have a nonlinearity that can be controlled by a DC applied voltage. Only for a small range of applied voltages is a shape-preserving spatial profile observed to propagate throughout the crystal. These solitons are independent of the light intensity and provide trapping in two dimensions. These are quasi-steady-state solitons, existing only in the time window between the formation of the space-charge grating and the screening of the applied field.

A second kind of photorefractive soliton, called the *screening soliton*, appears in steady state due to the nonuniform screening of the applied field because of nonuniform intensity distribution that can take place in photo-voltaic media (Segev, 1994). A third kind of photo-refractive soliton takes place in lithium niobate, which has a strong photovoltaic current. This results in a photo-voltaic field that changes the refractive index to enable a one-dimensional self-guided spatial soliton (Taya, 1995). The ease with which spatial solitons can now be created has opened up a huge field for experimental study (Stegeman, 1999) with creative new optical profiles (Shu, 2010). Applications are not so readily available, however.

9.2.2 Spark tracks in air

Laser-produced filamentary sparks are the result of instabilities in nonlinear media, particularly in air. The separate regions of ionization suggest that the spatial distribution of the electric field needed for ionization and created by the focused laser beam has regions of maximum and minimum intensities along the beam axis (Berge 2007). Laser-induced breakdown and resulting filaments in air is now a very large field, with work underway to use in weapons.

9.3 Nonlinear waveguides

Waveguides can increase the effective length over which laser light can propagate at high intensity down a nonlinear medium. They can also increase the interaction length between two light waves, enhancing nonlinear phenomena. In addition, there are some particular ways in which the nonlinear refractive index can create or destroy waveguides. Nonlinear waveguides can exhibit optical bistability (which will be discussed in detail later); this may include thermal nonlinearities as well as the usual χ_3 nonlinearities.

Waveguides increase the effective path length for NLO processes, such as SHG. These processes usually increase quadratically with the interaction length (at least until the process begins to saturate). But the process also depends on the intensity. To achieve high intensity, it is necessary to focus the beam, which usually results in an interaction length only twice the Rayleigh length. Waveguides are the way to overcome this limitation. Waveguides have become important for harmonic generation, because quasi-phase matching is often simpler to create in waveguides than in bulk. Path lengths go from tens of micrometers in a focused beam to several cm, making harmonic generation very practical even for mW lasers.

Highly nonlinear waveguides can be created or destroyed by intense incident light. This is a form of all-optical switching that has been investigated for integrated photonics. Optical creation of waveguides can be seen in bulk photorefractive media, when optical solitons are formed. The nonlinearity may be located in the waveguiding medium itself or in one or more of the media bounding the waveguide.

Within nonlinear waveguides, the shape of the guided mode and its propagation wavevector depend strongly on the optical power. This means incident light power can vary its own coupling efficiency into the waveguide, through the variation of the power-dependent nonlinear refractive index of the spatial layer in the coupling region. In a prism-coupling setup, the optimum coupling angle at high incident power is different from that at low incident power. In end-coupling into a waveguide, the input must have the exact mode shape to achieve (in principle) 100% coupling into the one mode in a single-mode guide. In a nonlinear waveguide, this mode-shape depends on intensity inside the guide. Because coupling into the waveguide depends on the intensity inside the guide, these nonlinear waveguides can exhibit *optical bistability.* By definition, optical bistability means that there are two possible output powers for a single input power. As the input intensity is turned up from zero, the coupling into the waveguide may be poor, because there may be mode mismatch. Thus the output power at a particular power, say P_o, would be some small fraction of the input power. As the input power is turned up toward its maximum, the intensity inside the waveguide may move it toward mode-match, so that a higher fraction of the incident light is coupled into the guide. Thus coupling into the waveguide is now high. Upon lowering the input power back to P_o, the waveguide remains closer to mode-match than it was when the input power was increased to P_o from below. Thus there are two possible output states when the input is P_o. This is an example of optical bistability arising from a non-local nonlinearity. The effective nonlinearity is non-local because the mode-shape arises from the waveguide definition of optimum mode shape. Potential applications lie in the area of all-optical signal processing: bistability, switching, upper and lower threshold devices, optical limiters.

Optical bistability has been observed as a result of opto-thermally-induced refractive index changes. Temperature-induced dispersive OB depends on the temperature change of the real part of the index of refraction affected by a change in the absorption coefficient via the Kramers-Kronig relation. The temperature-dependent change in the optical path length $\Delta(nL)$ is described by a total differential. Neglecting thermal expansion of the sample we have $\Delta(nL)=L\ (\partial n/\partial t)\Delta T$. This kind of nonlinearity requires feedback, forming a kind of NLFP.

Another kind of waveguide bistability utilizes the possibility that the absorption is nonlinear with temperature. Under the right conditions, this leads to a nonlinear equation that exhibits bistability in output vs. input. Semiconductors present particularly strong example of such bistability (Kim, 1988), as do a number of organic compounds. Polymer dispersed liquid crystals can show thermally induced optical bistability (Mormile, 1998).

Fibers offer an extraordinarily long path length in the χ_3 material fused silica. The ability to make a photonic fibes, with holes along the fiber (a photonic crystal fiber, or PCF) accurate located for specific applications, has made it possible to increase nonlinear effects. PCFs offer single-mode propagation over a broad wavelength range with better mode confinement, increasing the nonlinearity. It is also possible to engineer their group velocity dispersion so as to create phase match. Besides SRS and SBS and FWM, which have been already described in fibers, the phenomena of self-phase modulation and cross-phase modulation occur strongly in these fibers.

Air–silica microstructured optical fibers (sometimes called photonic bandgap crystal fibers) can exhibit anomalous dispersion at visible wavelengths. This provides the phase-matching necessary for a myriad of nonlinear interactions: spectral broadening and continuum generation, stimulated Raman and Brillouin scattering, and parametric amplification. Using photonic crystal fibers and 100 fs pulses, a supercontinuum can be generated, providing a light source from the infrared to the UV (Dudley, 2006). Supercontinuum can be generated using laser pulses as long as several ns or even with high power cw sources. Applications include optical coherence tomography, spectroscopy and optical frequency metrology, leading to the development of a new generation of optical clocks, which has opened up new perspectives to study limits on the drift of fundamental physical constants.

10. Cavity-enhanced nonlinearities

Another way to increase nonlinear effects is to place the nonlinear material in a reflective cavity to resonantly enhance the local optical field. The internal field is increased by the cavity Q. An obvious example is cavity-enhanced SHG. For some nonlinearities, the cavity can provide feedback so that an amplifying process becomes an oscillation. Examples are OPO's and Raman lasers. Finally, some phenomena require a cavity to be observed at all. The nonlinear Fabry-Perot demonstrates optical bistability, for example.

10.1 Resonantly enhanced wave-mixing

Placing a SHG crystal inside the pumping laser resonator has long been a way to increase the SHG efficiency. Internal intensities can be orders of magnitude larger than the external intensity, enabling much larger conversion efficiencies than placing the NLO crystal outside the resonator. The internal intensity $I_{inside} = I_{out}/(1-R)$, where R is the reflectivity of the output mirror, which can be seen by recognizing that (1-R) is the transmission T and I_{out} = T•I_{inside}. As an example, today's green laser pointers are frequency-doubled diode-pumped solid-state lasers with a cavity doubling crystal internal to the laser cavity. An alternative is to place the doubling crystal in a cavity *external* to the laser cavity. Conversion efficiencies as high as 75% have been reported with 60 mW cw output (Li 2006) with QPM based on periodic poling.

A cavity can also resonate four-wave mixing, to enhance the output. Examples are photo-refractive films and polymer films.

10.2 Optical Parametric Oscillators (OPOs)

An OPO converts monochromatic laser emission (the *pump*) into a tunable output via a three-wave mixing process. Quantum efficiencies can exceed 50%. The heart of an OPO is a nonlinear-optical (NLO) crystal characterized by an NLO coefficient, d_{eff} and its related NLO figure of merit, d_{eff}^2/n^3 (where n is the refractive index). In the NLO crystal, the pump photon decays into two less-energetic photons (the signal and the idler) so that the sum of their energies equals that of the pump photon. An important further constraint is that the sum of the signal and idler *wave-vectors* must equal that of the pump ("phase-matching" condition). The latter condition is never satisfied in the transparency range of isotropic media but can be fulfilled in birefringent crystals. Alternatively, it can be fulfilled in quasi-phase-matched (QPM) crystals with periodically modulated nonlinearity (periodically poled lithium niobate, for example) in which the artificially created grating compensates for the wave-vector mismatch.

Parametric frequency down-conversion in an OPO can be regarded as the inverse process of sum-frequency generation. Alternatively, an NLO crystal can be viewed simply as the impetus for the pump photon into break up into two smaller photons. Rotating the crystal changes the ratio between the signal and idler photon energies, and thus tunes the frequency of the output. The easiest way to illustrate parametric frequency conversion is to consider the case of a short (<1 ns) intense pulse as the pump. In this case, a single pass through an NLO crystal is sufficient to convert a substantial fraction of the pump into the signal and the idler. This type of single-pass device is called an optical parametric generator (OPG). For pump pulses with lower intensity, parametric frequency conversion is weaker; therefore, an OPO cavity is required to enhance this process.

The main value in OPOs is that the signal and idler wavelengths, which are determined by phase-matching, can be varied over a wide range. Thus it is possible to produce wavelengths which are difficult or impossible to obtain from any laser (e.g. in the mid-infrared, far-infrared or terahertz regions), with wide wavelength tunability.

10.2.1 Optical parametric oscillator threshold

The threshold for an OPO is calculated by equating the gain of the optical parametric amplifier to the losses in the cavity. The gain can be found by analyzing coupled mode theory for the mixing of the three waves. A strong input pump field E_3 at frequency ω_3 and a weak signal field E_2 at frequency ω_2, which the parametric process will amplify, are incident in a nonlinear medium. The parametric process invokes an idler field at frequency E_1 and frequency ω_1 to complete the interaction, assumed here to be phase-matched: $k_1 + k_2 = k_3$. The coupled mode approach provides three equations:

$$\frac{dE_1(z)}{dz} = i\kappa_1 E_3(z)E_2^*(z) \qquad \frac{dE_2(z)}{dz} = i\kappa_2 E_3(z)E_1^*(z) \qquad \frac{dE_3(z)}{dz} = i\kappa_3 E_1(z)E_2(z)$$

Assuming no input idler light and no pump depletion, the signal intensity increases with length ℓ as $I_2(\ell) = I_2(0) \sinh^2(\Gamma\ell)$,

where I is the intensity in W/m^2; Δk is the phase mismatch; and Γ is the gain factor defined as

$$\Gamma^2 = \frac{8\pi^2 d_{\text{eff}}^2}{c\varepsilon_0 n_1 n_2 n_3 \lambda_1 \lambda_2} I_3(0)$$

Under these assumptions, the signal intensity gain per unit length has a simple form $G_2(\ell) = \sinh^2(\Gamma\ell)$, which for small gains increases quadratically and for large gains increases exponentially as $\exp(2\Gamma\ell)$.

The threshold for the OPO is found by setting the signal gain per unit length equal to its resonator loss per unit length, α, given by $\exp(2\alpha\ell) = R_1 R_2$, to obtain $I_{th} = \alpha^2(c\varepsilon_0 n_1 n_2 n_3 \lambda_1 \lambda_2)/[8(\pi\ell d_{\text{eff}})^2]$. The threshold intensity decreases as the resonator loss decreases (mirror reflectances increase), and quadratically as the length and nonlinearity increase.

10.2.2 OPO applications

OPOs are useful sources for high peak or average power, high conversion efficiency, and broad continuous tunability. They are particularly valuable in the mid-IR (wavelengths >2.5 μm) where there are no tunable lasers similar to Ti:sapphire. New nonlinear-optical materials have enabled compact and efficient OPO's with infrared wavelength tunability far beyond 5 μm, opening up new applications in molecular spectroscopy, atmospheric monitoring, and ultra-sensitive detection. In the 2- to 20-μm portion of the spectrum, gases exhibit uniquely identifiable absorption features. Pollution monitoring, atmospheric chemistry, and chemical and biological warfare detection can benefit from compact and efficient mid-IR laser sources that allow detection of trace gases and vapors by volume, down to the part-per-billion level. Other applications include noninvasive medical diagnosis by breath analysis, ultrasensitive detection of drugs and explosives down to the parts-per-trillion level using cavity ring-down spectroscopy, and short-range terrestrial or near-earth communications.

The OPO is widely used to generate squeezed coherent states and entangled states of light. Considering a single photon in the OPO, each pump photon gives rise to a pair of photons; the signal and idler fields are correlated at the quantum level, which is required for squeezing. The phases of the signal and idler are correlated as well, leading to entanglement, which is a key requirement for quantum computing.

10.3 Nonlinear resonators: Fabry-Perots and rings

When feedback is added to a nonlinear refractive index or absorption interaction, optical switching, optical bistability and multistability can occur, with potential for all-optical logic and computing. While tantalizing for practical applications, the NLFP (nonlinear Fabry-Perot) has rarely seen practicality. In photonics, the NLR (nonlinear ring) may play more of a role. The feedback effect of a cavity can also be artificially created in a hybrid electrical-optical device, which may have their own applications.

Understanding the origin of optical bistability in a NLFP is straight-forward. It was first described using saturable absorption and later it was realized that bistability could be achieved more with a nonlinear refractive index. Suppose a resonator has a characteristic transmission of $T_r = I_{out}/I_{in} = T_r(I_{inside})$, where the last equality defines the functional form of the resonator transmission as a function of the nonlinear absorption or refractive index, which itself depends on intensity. Note the three different values of the intensity: input intensity I_{in}, output intensity I_{out}, and intensity inside the resonator, I_{inside}. What makes the nonlinear resonator unique is that the light intensity inside the resonator does *not* depend directly on the input light intensity. Instead, the intensity coming *out* of the resonator is proportional to the intensity *inside*. That is, $I_{out} = T_o I_{inside}$, where T_o is the transmission of the output port of the resonator. In a Fabry-Perot, $T_o = 1-R$, where R is the reflectivity of the output mirror. Thus the resonator obeys the following: $T_r(I_{out}/T_o) = I_{out}/I_{in}$. This can be evaluated through plotting $I_{in} = I_{out}/T_r(I_{out}/T_o)$. If the functional form of T_r is multi-valued, such as in a refractive nonlinear Fabry-Perot, whose transmission values repeat modulo 2π, plotting the output vs. the input may demonstrate optical bistability or multi-stability.

Assume a saturable absorption of amount $\Delta\alpha$, lying on a base of unsaturable absorption, with a form given by $\alpha = \alpha_B + \Delta\alpha/(1 + I/I_s)$. It can be shown that to observe bistability it is necessary that $\alpha L/(T + \alpha_B L) > 8$, where T is the transmission of the lossless cavity. Even if $T \to 0$, bistability still requires that $\Delta\alpha > 7\alpha_B$. Nonlinearities are usually not this large, and so saturable absorption bistability is not usually the predominant form.

Bistability in a NLFP is usually due to a nonlinear refractive index. When the nonlinearity is saturating, it can be shown that if Δn is the maximum refractive index change, then the condition for bistability is $\Delta n k_o L > \alpha L + (1-R_{eff})$, where α is the (unsaturable) loss per unit length inside the cavity and R_{eff} is the average mirror reflectivity (Garmire, 1989). Semiconducting quantum wells have been shown to exhibit optical bistability when placed in a Fabry-Perot (Vivero, 2010), as does porous silicon (Pham, 2011).

Laser diodes also exhibit optical bistability. Many suggestions have been made to make practical all-optical switching devices in semiconductor materials, but to date few practical systems have arisen. One important application for resonating saturable losses has arisen, however. Bragg mirrors with semiconductor saturable absorbers have been shown to be excellent mode-locking devices for femtosecond lasers (Keller, 1996; Khadour, 2010).

The ring resonator is analogous to a Fabry-Perot and therefore exhibits bistability. These resonators are used in optical waveguide circuits and have become more and more practical as technology for micro-circuits has improved. Optical bistability occurs in a 5 μm radius ring resonator fabricated on a highly integrated silicon device. Strong light-confinement makes possible a nonlinear optical response in silicon with pump power of 45 mW, due to a thermal nonlinearity. In such a small device, the thermal speeds can be up to 500 kHz (Almeida, 2004). The ring resonator directionally coupled to a channel waveguide forms a wavelength-sensitive add-drop filter that is very useful in WDM optical communications. The nonlinear ring enables all-optical switching and logic (Parisa, 2009). Photonic crystals can be engineered to replace optical waveguides and provide even more light confinement and even lower switching powers (Yanik, 2003). While still in the realm of research, nonlinear photonic crystal devices are expected to have a bright future in photonics.

10.4 Hybrid optical bistability

Optical bistability can be created by use of electrical feedback, forming a hybrid system. When some fraction of the output of an optical modulator is incident on a photo-detector, its electrical output can be fed back as a change in voltage on the modulator, changing its transmission (Garmire, 1978). This leads to a hybrid bistability that opens up the number of devices that can exhibit optical bistability. One example is the development of devices that might allow parallel interconnects in complex computer systems (Miller, 2000).

11. NLO topics not covered

This section briefly mentions a number of topics in NLO that were not covered in this chapter. Detailed applications of nonlinearities have been skipped, as well as some recent cutting edge research. Theoretical work and modeling has been crucial to the development of NLO, mostly not discussed here. Indeed, inventive ideas, high quality experiments, physically intuitive modeling and deep theoretical understanding have all played their part in creating this exciting field of NLO.

The laser is an exceedingly nonlinear device and, as such, can be used to demonstrate many of the topics discussed in this chapter. Nonlinear gain can be considered an analog to nonlinear absorption; in semiconductor lasers, a change in gain also changes the refractive index. Optical nonlinearities are needed to explain much of the behavior of semiconductor lasers. Mode-locking and Q-switching often use nonlinear media inserted into the laser cavity, and their interaction with the nonlinearities within the laser must be understood. This chapter is limited to passive devices, not including lasers.

This chapter treats mostly stable nonlinear device performance. However, spatial and temporal instabilities can easily arise in nonlinear systems (Cross, 1993), leading to spatial multi-filamentation, as well as self-organization and even chaos. In the time domain, self-pulsing and chaos can also occur, particularly if light is fed back into the nonlinear medium with a time delay (Goldstone, 1983). Such nonlinearities are seen particularly in lasers (Blaaberg, 2007) and nonlinear fibers (Kibler, 2010).

Plasmons are created at the surface between a dielectric and a metal film, due to interaction between propagating light and the metal; interesting phenomena occur if the dielectric medium is nonlinear. Polaritons result from strong coupling between light and an excited electric dipole and can lead to optical nonlinearities. Neither are approached here.

NLO has revolutionized spectroscopy, which now has a vast number of applications, in chemistry, in biology, in environmental studies, etc. For example, nonlinear saturation enables spectroscopists to make measurements inside inhomogeneously broadened lines; multi-photon absorption enables measurement of levels that are symmetry-forbidden in usual one-photon spectroscopy. This chapter touches only on a few examples.

NLO has revolutionized other fields of science. For example, SHG and OPOs provide sources for squeezed light, cooling atoms and molecules to achieve Bose-Einstein condensation, etc. NLO has made ultra-fast optics possible, exploring ultra-fast processes in molecules, solid state materials, chemistry, plasma physics, etc. Nonlinear frequency conversion has enabled ultra-stable frequency sources that have become new, highly-accurate standards, and led to two Nobel prizes. Two-photon fluorescence and Raman lasers are just two examples of techniques that are standard in biomedical research.

The topic categories covered at the 2011 Nonlinear Optics Conference, sponsored by the Optical Society of America give a good idea of the breadth of non-linear optics and what is cutting-edge NLO research. Fundamental studies and new concepts: Quantum optics, computation and communication; single-photon nonlinear optics; Solitons and nonlinear propagation; Ultrafast phenomena and techniques; Surface, interface and nanostructure nonlinearities; Microcavity and microstructure phenomena; High intensity and relativistic nonlinear optics; Slow light; Coherent control; Pattern formation in nonlinear optical systems. Nonlinear media investigated today are: Atoms, molecules and condensates; Cold atoms; Dielectrics; Semiconductors; Nanostructures; Photonic bandgap structures; Fibers and waveguides; Photorefractives; Nonlinear nanophotonics. Key areas that merge science and applications are: Novel lasers and frequency converters; Micro solid-state photonics. Applications of interest to the NLO community today include: Lasers and amplifiers; Frequency converters and high harmonics generation; Optical communications; Photonic switching; Ultrafast measurement; Nonlinear x-ray optics; Materials processing; Optical storage; Biological elements; Laser induced fusion; Frequency combs and optical clocks.

12. Conclusions

Nonlinear Optics, has been described here in mostly classical terms. Traditional second harmonic generation, sum-frequency and difference-frequency generation, and generation of a DC field take place only in transparent media that lack a center of inversion symmetry and require phase-matching – with an anisotropic crystal or with periodic poling, or other means of quasi-phase matching. Third order nonlinearities do not necessarily require a center of inversion symmetry, nor phase-matching, although these may be required by some processes. These processes may be enhanced by proximity of an atomic or molecular resonance, although these are not required. These NLO processes are described by a dielectric susceptibility that depends on the light's electric field. Closely related are parametric processes, such as the optical parametric oscillator and the optical parametric amplifier. Nonlinear absorption processes may either cause a decrease in a strong absorption line or may increase absorption due to multi-photon processes. Stimulated Raman and Brillouin scattering are laser-like manifestations of well-known low-power phenomena. The former offers an array of new wavelengths, the latter enables phase conjugation.

This review shows how vast the field of nonlinear optics is today and how far it has come since second harmonic generation was demonstrated with a ruby laser in 1961.

13. References

Abeeluck, A.K. & Garmire, E. (2002). Diffraction Response of a Low-Temperature-Grown Photorefractive Multiple Quantum Well Modulator. *Journal of Applied Physics*, Vol. 91, No. 5, March 2002, pp. 2578-2586.

Abolghasem, P., Han, J., Bijlani, B.J., Helmy, A.S. (2010). Type-0 second order nonlinear interaction in monolithic waveguides of isotropic semiconductors. *Optics Express*, Vol. 18, No. 12, Jun 2010, pp. 12681-12689.

Agrawal, G.P. (2010). *Fiber-Optic Communication Systems* (4th ed). Wiley, ISBN: 978-0-470-50511-3, Hoboken, NJ.

Aitchison, J.S., Weiner, A.M., Silberberg, Y., Oliver, M.K., Jackel, J.L., Leaird, D.E., Vogel, E.M. & Smith, P.W.E. (1990). Observation of Spatial Optical Solitons in a Nonlinear Glass Waveguide. *Optics Letters*, Vol. 15, No. 9, May 1990, pp. 471-473.

Almeida, V.R., Lipson, M. (2004). Optical bistability on a silicon chip. *Optics Letters*, Vol. 29, No. 20, Oct 2004, pp. 2387-2389.

Armstrong, J.A., Bloembergen, N., Ducuing, J. & Pershan, P.S. (1962). Interactions between Light Waves in a Nonlinear Dielectric. *Physical Review*, Vol. 127, No. 6, Sept 1962, pp. 1918-1939.

Baehr-Jones, T., Hochberg, M., Wang, G., Lawson, R., Liao, Y., Sullivan, P., Dalton, L., Jen, A., & Scherer, A. (2005). Optical modulation and detection in slotted Silicon waveguides. *Optics Express*, Vol. 13, No. 14, July 2005, pp. 5216-5226.

Bao X., Chen L. (2011). Recent Progress in Brillouin Scattering Based Fiber Sensors, *Sensors*, Vol. 11, No. 4, Apr. 2011, pp. 4152-4187.

Barthelemy, A., Maneuf, S. & Froehly, C. (1985). Propagation Soliton Et Auto-Confinement De Faisceaux Laser Par Non Linearite Optique De Kerr. *Optics Communications*, Vol. 55, No. 3, September 1985, pp. 202-206.

Becerra, F.E., Willis, R.T., Rolston, S.L., Carmichael, H. J., & L. A. Orozco, L.A. (2010). Nondegenerate four-wave mixing in rubidium vapor: Transient regime, *Physical Review A* Vol. 82, Oct 2010, pp. 043833,1-9.

Begley, R.F., Harvey, A.B. & Byer, R.L. (1974). Coherent Anti-Stoke Raman Spectroscopy. *Applied Physics Letters*, Vol. 25, No. 7, Oct 1974, pp. 387-390.

Berge, L., Skupin, S., Nuter, R., Kasparian, J., & Wolf, J.P. (2007). Ultrashort Filaments of Light in Weakly Ionized, Optically Transparent Media. *Reports on Progress in Physics*, Vol. 70, No. 10, Sept 2007, pp. 1633-1713.

Bergmann, K., Theuer, H., & Shore, B.W. (1998). Coherent Population Transfer Among Quantum States of Atoms and Molecules. *Review Modern Physis*, Vol. 70, No. 3, July 1998, pp. 1003-1025.

Blaaberg, S., Petersen, P.M. & Tromborg, B. (2007) Structure, stability, and spectra of lateral modes of a broad-area semiconductor laser, *IEEE Journal Of Quantum Electronics*, Vol. 43, No. 11-12, Nov-Dec 2007, pp. 959-973.

Blanche, P.-A., Bablumian, A., Voorakaranam, R., Christenson, C., Lin, W., Gu, T., Flores, D., Wang, P., Hsieh, W.-Y., Kathaperumal, M., Rachwal, B., Siddiqui, O., Thomas, J., Norwood, R.A., Yamamoto, M. & Peyghambarian, N. (2010). Holographic Three-Dimensional Telepresence using Large-Area Photorefractive Polymer. *Nature*, Vol. 468, Nov 2010, pp. 80–83.

Bloembergen, N. (1967). Stimulated Raman Effect. American Journal of Physics, Vol. 35, No.11, Nov 1967, pp. 989-1023.

Bondani, M. & Andreoni, A. (2002). Holographic Nature of Three-Wave Mixing. *Physical Review A*, Vol. 66, No. 3, Sept 2002, pp. 033805 (1-9).

Bowers, M.W., Boyd, R.W. & Hankla, A.K. (1997). Brillouin-Enhanced Four-Wave-Mixing Vector Phase-Conjugate Mirror with Beam-Combining Capability. *Optics Letters*, Vol. 22, No. 6, March 1997, pp. 360-362.

Brewer, R.G. Lifsitz, J.R. Garmire, E. Chiao, R.Y. & Townes, C.H. (1968). Small-Scale Trapped Filaments in Intense Laser Beams. *Physical Review*, Vol. 166, No. 2, Feb 1968, pp. 326-331.

Broderick, N.G.R., Ross, G.W., Offerhaus, H.L., Richardson, D.J. & Hanna, D.C. (2000). Hexagonally poled lithium niobate: A two-dimensional nonlinear photonic crystal. *Physical Review Letters*, Vol. 84, No. 19, May 2000, pp. 4345-4348.

Byer, R.L. (1997). Quasi-phasematched nonlinear interactions and devices. *Journal of Nonlinear Optical Physics and Materials*, Vol. 6, No. 4, Dec 1997, pp. 549-592.

Chang, T.Y. & Hellwarth, R.W. (1985). Optical-Phase Conjugation by Backscattering in Barium-Titanate. Optics Letters, Vol. 10, No. 8, Aug 1985, pp. 408-410.

Chen, D., Qin, S. & He, S. (2007). Channel-Spacing-Tunable Multi-Wavelength Fiber Ring Laser with Hybrid Raman and Erbium-doped Fiber Gains. *Optics Express*, Vol. 15, No. 3, Feb 2007, pp. 930-935.

Chi, S.H., Hales, J.M., Cozzuol, M., Ochoa, C., Fitzpatrick, M. & Perry, J.W. (2009), Conjugated polymer-fullerene blend with strong optical limiting in the near-infrared, *Optics Express*, Vol. 17, No.24, Dec 2009, pp. 22062-22072.

Chiao, R.Y., Garmire, E. & Townes, C.H. (1964). Self-Trapping of Optical Beams. *Physical Review Letters*, Vol. 13, No. 15, Oct 1964, pps. 479-482.

Cronin-Golomb, M., Fischer, B., White, J.O. & Yariv, A. (1984). Theory and Applications of 4-Wave Mixing In Photorefractive Media. *IEEE Journal of Quantum Electronics*, Vol. 20, No. 1, Jan 1984, pp. 12-30.

Cross M.C. & Hohenberg, P.C. (1993). Pattern-Formation Outside of Equilibrium, *Reviews of Modern Physics*, Vol. 65, No. 3, Jul 1993, pp. 851-1112.

Dainese, P., Russell, P. St. J., Joly, N., Knight, J.C., Wiederhecker, G.S., Fragnito, H.L., Laude, V. & Khelif, A. (2006). Stimulated Brillouin Scattering from Multi-GHz-Guided Acoustic Phonons in Nanostructured Photonic Crystal Fibres. *Nature Physics*, Vol. 2, May 2006, pp. 388-392.

Damzen, M.J., Vlad, V., Mocofanescu, A. & Babin, V. (2003). *Stimulated Brillouin Scattering: Fundamentals and Applications (Series in Optics and Optoelectronics)*. CRC Press, 2003, ISBN 0750308702, Institute of Physics, London.

Dane, C.B., Neuman, W.A. & Hackel, L.A. (1994). High-Energy SBS Pulse-Compression. *IEEE Journal of Quantum Electron*, Vol. 30, No. 8, Aug 1994, pp. 1907-1915.

Demtröder, D. (2008). *Laser Spectroscopy: Vol. 2: Experimental Techniques* (4th ed). ISBN-10: 3540749527, ISBN-13: 978-3540749523, Springer-Verlag, Berlin.

DeSalvo, R., Said, A.A., Hagan, D.J., Van Stryland, E. W., & Sheik-Bahae M. (1996). Infrared to Ultraviolet Measurements of Two-Photon Absorption and n_2 in Wide Bandgap Solids. *IEEE Journal Quantum Electron*, Vol. 32, No. 8, Aug 1996, pp. 1324-1333.

Dohler, G.H. (1986). Doping Superlattices (n-i-p-i Crystals). *IEEE Journal of Quantum Electronics*, Vol. 22, No. 9, Sept 1986, pp. 1682-1695.

Dragoman, D. & Dragoman, M. (2004). Terahertz fields and applications. *Progress in Quantum Electronics*, Vol. 28, No. 1, Jan 2004, pp. 1-66.

Dudley, J.M., Genty, G. & Coen, S. (2006). Supercontinuum generation in photonic crystal fiber. Review of Modern Physics, Vol. 78, No. 4, Oct-Dec 2006, pp. 1135-1184.

Duncan, M.D., Mahon, R., Tankersley, L.L. & Reintjes, J. (1988). Transient Stimulated Raman Amplification in Hydrogen. *Journal of the Optical Society of AmericaB-Optical Physics*, Vol. 5, No. 1, Jan 1988 (A) , pp. 37-52.

Duree, G.C., Schultz, J.L., Salamo, G.J., Segev, M., Yariv, A., Crosignani, B., DePorto, P., Sharp, E.J. & Neurgaonkar, R.R. (1993). Observation of Self-Trapping of an Optical Beam due to the Photorefractive Effect. *Physical Review Letters*, Vol. 71, No. 4, July 1993, pp. 533-536.

Eralp, M., Thomas, J., Tay, S., Li, G., Schülzgen, A., Norwood, R.A., Yamamoto, M. & Peyghambarian, N., "Submillisecond response of a photorefractive polymer under single nanosecond pulse exposure," (2006). *Applied Physics Letters*, Vol. 89, No.11, November, 2006 pp. 114105-114107.

Erokhin, A.I. & Smetanin, I.V. (2010). Experimental and Theoretical Study of Self-Phase Modulation in SBS Compression of High-Power Laser Pulses, *Journal of Russian Laser Research*, Vol. 31, No. 5, Sept 2010, pp. 452-461.

Feinberg, J., Heiman, D., Tanguay, A.R. & Hellwarth, R.W. (1980). Photorefractive Effects and Light-Induced Charge Migration in Barium-Titanate. Journal of Applied Physics, Vol. 51, No. 3, March 1980, pp. 1297-1305.

Fiebig, M., Lottermoser, T., Frohlich, D., Goltsev, A.V., & Pisarev, R.V. (2002). Observation of coupled magnetic and electric domains. *Nature*, Vol. 419, No. 6909, Oct 2002, pp. 818-820.

Fiore, A., Berger, V., Rosencher, E., Bravetti, P. & Nagle, J. (1998). Phase matching using an isotropic nonlinear optical material, *Nature*, Vol. 391, Jan 1998, pp. 463-466.

Fleischhauer, M., Imamoglu, A. & Marangos, J.P. (2005). Electromagnetically Induced Transparency: Optics in Coherent Media. *Reviews of Modern Physics*, Vol. 77, No. 2, April 2005, pp. 633 - 673.

Franken, P.A., Hill, A.E., Peters, C.W. & Weinreich, G. (1961). Generation Of Optical Harmonics. *Physical Review Letters*, Vol. 7, No. 4, Aug 1961, pp.118-120.

Fueloep, J.A., Palfalvi L., Hoffmann M.C. & (Hebling, Janos), Towards generation of mJ-level ultrashort THz pulses by optical rectification, *Optics Express* Vol. 19, No. 16, Aug. 2011, pp. 15090-15097.

Garmire, E. (1989). Criteria For Optical Bistability In A Lossy Saturating Fabry-Perot. *IEEE Journal of Quantum Electronics*, Vol. 25, No. 3, Mar 1989, pp. 289-295.

Garmire, E. & Townes, C.H. (1964). Stimulated Brillouin Scattering in Liquids. *Applied Physics Letters*, Vol. 5, No. 4, Aug 1964, pp. 84.

Garmire, E., Chiao, R.Y. & Townes, C.H. (1966) Dynamics and Characteristics of the Self-Trapping of Intense Light Beams. *Physical Review Letters*, Vol. 16, No. 9, Feb 1966, pp. 347-349.

Garmire, E., Jokerst, N.M., Kost, A., Danner, A. & Dapkus, P.D. (1989). Optical Nonlinearities Due to Carrier Transport in Semiconductors. *Journal of the Optical Society of America B-Optical Physics*, Vol. 6, No 4, April 1989, pp. 579-587.

Garmire, E., Marburger, J.H., Allen, S.D. (1978). Incoherent Mirrorless Bistable Optical-Devices. Applied Physics Letters, Vol. 32, No. 5, March 1978, pp. 320-321.

Garmire, E., Townes, C.H. & Pandarese, F. (1963). Coherently Driven Molecular Vibrations and Light Modulation. *Physical Review Letters*, Vol. 11, No. 4, Aug 1963, pp. 160-163.

Geng, J., Staines, S., Wang, A., Zong, J., Blake, M. & Jiang, S. (2006). Highly Stable Low-Noise Brillouin Fiber Laser with Ultranarrow Spectral Linewidth. *IEEE Photonics Technology Letters*, Vol. 18, No. 17, Sept 2006, pp. 1813-1815.

Genty, G., Coen, S., & Dudley, J. M. (2007) Fiber supercontinuum sources (Invited). *Journal of the Optical Society of America B-Optical Physics*, Vol. 24, No. 8, Aug 2007, pp. 1771-1785.

Goldstone, J.A. & Garmire, E.M. (1983). Regenerative Oscillation in the Non-Linear Fabry-Perot-Interferometer. *IEEE Journal of Quantum Electronics*, Vol. 19, No. 2, Feb 1983, pp. 208-217.

Gontier, Y. & Trahin, M. (1968). Multiphoton Ionization of Atomic Hydrogen in Ground State. *Physical Review*, Vol. 172, No. 1, Aug 1968, pp. 83-87.

Grasiuk, A.Z., Kurbasov, S.V. & Losev, L.L. (2004). Picosecond Parametric Raman Laser Based on KGd(WO$_4$)$_2$ Crystal. *Optics Communications*, Vol. 240, No. 4-6, Oct 2004, pp. 239-244.

Grant-Jacob, J., Mills, B., Butcher, T.J., Chapman, R.T., Brocklesby, W.S. & Frey, J.G. (2011) Gas jet structure influence on high harmonic generation, *Optics Express.* Vol.19, No.10, May 2011, pp.9801-9806

Grazulevicius, J.V., Strohriegl, P., Pielichowski, J. & Pielichowski, A. (2003). Carbazole-containing polymers: synthesis, properties and applications. *Progress in Polymer Science,* Vol. 28, No. 9, Sept 2003, pp. 1297-1353.

Guo, J., Togami, T., Benten, H., Ohkita, H. & Shinzaburo Ito, S. (2009) Simultaneous multi-photon ionization of aromatic molecules in polymer solids with ultrashort pulsed lasers *Chemical Physics Letters,* Vol. 475, No. 4-6, Jun 2009, pp. 240-244.

Gunter, P. (1982). Holography, Coherent Light Amplification and Optical Phase Conjugation with Photorefractive Materials. *Physics Reports (Review Section of Physics Letters),* Vol. 93, No. 4, Dec 1982, pp. 199-299.

Hagley, E.W., Deng, L., Kozuma, M., Wen, J., Helmerson, K.S., Rolston, L. & Phillips, W.D. (1999). A Well-Collimated Quasi-Continuous Atom Laser. *Science,* Vol. 283, No. 5408, March 1999, pp. 1706-1709.

Haus, H.A. & Wong, W.S. (1996). Solitons in Optical Communications. Reviews of Modern Physics, Vol. 68, No. 2, April 1996, pp. 423-444.

Headley, C. & Agrawal, G.P. (Eds.) (2005). *Raman Amplification in Fiber Optical Communication Systems.* Academic Press, ISBN 10: 0-12-044506-9, ISBN 13: 978-0-12-044506-6, Elsevier, Burlington MA.

Heebner, J.E., Bennink, R.S., Boyd, R.W. & Fisher, R.A. (2000). Conversion of Unpolarized Light to Polarized Light with Greater than 50% Efficiency by Photorefractive Two-Beam Coupling. *Optics Letters,* Vol. 25, No. 4, Feb 2000, pp. 257-259.

Hellwarth, R.W. (1977). Third-Order Optical Susceptibilities of Liquids and Solids. *Monographs: Progress in Quantum Electronics,* J.H. Sanders and S. Stenholm, eds., Pergamon Press, New York, 1977, Part I, Vol. 5, pp. 1-68.

Hon, D.T. (1980). Pulse-Compression by Stimulated Brillouin-Scattering. *Optics Letters,* Vol. 5, No. 12, Dec 1980, pp. 516-18.

Horiguchi, T., Shimizu, K., Kurashima, T., Tateda, M., & Koyamada, Y. (1995). Development of a Distributed Sensing Technique Using Brillouin-Scattering. *Journal of Lightwave Technology,* Vol. 13, No. 7, Jul 1995, pp. 1296-1302.

Hsu, C.J., Chiu, K.P., Lue, J.T. & Lo, K.Y. (2011), Enhancement of reflective second harmonic generation using periodically arrayed silver-island films, *Applied Physics Letters,* Vol. 98, No. 26, Jun 2011, pp. 261107,1-3.

Hung, L.J., Diep, L.N., Han, C.C., Lin, C.Y., Rieger, G.W., Young, J.F., Chien, F.S.S. & Hsu, C.C. (2008). Fabrication of spatial modulated second order nonlinear structures and quasi-phase matched second harmonic generation in a poled azo-copolymer planar waveguide, *Optics Express,* Vol. 16, No.11, May. 2008 pp. 7832-7841.

Islam, M.N. (ed.) (2004). *Raman Amplifiers for Telecommunications 1: Physical Principles.* Springer Series in Optical Sciences, Springer-Verlag, ISBN-10: 1441918396, ISBN-13: 978-1441918390, NYC, NY.

Jansen, S.L., van den Borne, D., Spinnler, B., Calabro, S., Suche, H., Krummrich, P.M., Sohler, W., Khoe, G.-D. & de Waardt, H. (2006). Optical phase conjugation for ultra long-haul phase-shift-keyed transmission. *Journal of Lightwave Technology,* Vol. 24, No. 1, Jan. 2006, pp. 54- 64.

Jokerst, N.M. & Garmire E. (1988). Nonlinear Optical-Absorption In Semiconductor Epitaxial Depletion Regions. *Applied Physics Letters*, Vol. 53, No. 10, Sept 1988, pp. 897-899

Kaiser, W. & Garrett, C.G.B. (1961). 2-Photon Excitation in CaF_2 - Eu^{2+}. *Physical Review Letters*, Vol. 7, No. 6, Sept 1961, pp. 229-231.

Kamanina, N.V., Reshak, A.H., Vasilyev, P.Ya., Vangonen, A.I., Studeonov, V.I., Usanov, Yu.E., Ebothe, J., Gondek, E. , Wojcik, W. & Danel A. (2009). "Nonlinear absorption of fullerene- and nanotubes-doped liquid crystal systems", *Physica E*, Vol. 41, No. 1, Jan 2009, pp.391-394.

Keller, U., Weingarten, K.J., Kartner, F.X., Kopf, D., Braun, B., Jung, I.D., Fluck, R., Honninger, C., Matuschek, N. & derAu, J.A. (1996). Semiconductor Saturable Absorber Mirrors (SESAM's) for Femtosecond to Nanosecond Pulse Generation in Solid-State Lasers. *IEEE Journal of Selected Topics in Quantum Electronics*, Vol. 2, No. 3, Sept 1996, pp. 435-453.

Kim, B.G., Garmire, E., Shibata, N., & Zembutsu, S. (1987). Optical Bistability and Nonlinear Switching Due to Increasing Absorption in Single-Crystal Znse Wave-Guides. *Applied Physics Letters*, Vol. 51, No. 7, Aug 1987, pp. 475-477.

Kim, E.J., Yang, H.R., Lee, S.J., Kim, G.Y., Lee, J.W. & Kwak, C.H. (2010) Two beam coupling gain enhancements in porphyrin:Zn-doped nematic liquid crystals by using grating translation technique with an applied dc field, *Optics Communications*, Vol. 283, No. 7, Apr 2010, pp. 1495-1499.

Kirkwood,R.K., Michel, P., London, R. et al. (2011). Multi-beam effects on backscatter and its saturation in experiments with conditions relevant to ignition, *Physics of Plasmas* Vol. 18, May 2011, pp. 056311,1-13.

Khadour, A., Bouchoule, S., Aubin, G., Harmand, J.C., Decobert, J. & Oudar, J.L. (2010). Ultrashort pulse generation from 1.56 μm mode-locked VECSEL at room temperature, *Optics Express* Vol. 18, Sept 2010, pp. 19902-19913.

Kobyakov, A., Kumar, S., Chowdhury, D.Q., Ruffin, A.B., Sauer, M., Bickham, S.R. & Mishra, R. (2005). Design concept for optical fibers with enhanced SBS threshold. *Optics Express*, Vol. 13, No. 14, Jul 2005, pp. 5338-5346.

Kost, A., Garmire, E., Danner, A. & Dapkus, P.D. (1988). Large Optical Nonlinearities in a GaAs/AlGaAs Hetero n-i-p-i Structure. *Applied Physics Letters*, Vol. 52, No. 8, Feb 1988, pp. 637-639.

Kovalev, V.I., Harrison, R.G. & Scott, A.M. (2005). 300 W quasi-continuous-wave diffraction-limited output from a diode-pumped Nd : YAG master oscillator power amplifier with fiber phase-conjugate stimulated Brillouin scattering mirror. *Optics Letters* , Vol. 30, No. 24, Dec 2005, pp. 3386-3388.

Kowalczyk, T.C., Singer, K.D. & Cahill, P.A. (1995). Anomalous-Dispersion Phase-Matched 2nd-Harmonic Generation In A Polymer Wave-Guide. *Optics Letters*, Vol. 20, No. 22, Nov 1005, pp. 2273-2275.

Lenner, M., Rácz, P., Dombi, P., Farkas, G. & Kroó N. (2011). Field enhancement and rectification of surface plasmons detected by scanning tunneling microscopy. *Physical Review B*, Vol. 83, No. 20, May 2011, pp. 205428(1-5).

Lee K.J., Liu S., Gallo K., Petropoulos, P. & Richardson, D.J. (2011) Analysis of acceptable spectral windows of quadratic cascaded nonlinear processes in a periodically poled lithium niobate waveguide, *Optics Express*, Vol. 19, No. 9, Apr. 2011, pp. 8327-8335.

Li, M.J., Demichell, M., He, Q. & Ostrowsky, D.B. (1990). Cerenkov Configuration Second-Harmonic Generation In Proton-Exchanged Lithium-Niobate Guides. *IEEE Journal of Quantum Electronics*, Vol. 26, No. 8, Aug 1990, pp. 1384-1393.

Li, S., Zhang, X., Wang, Q., Zhang, X., Cong, Z., Zhang, H. & Wang, J. (2007). Diode-Side-Pumped Intracavity Frequency-Doubled Nd:YAG/BaWO$_4$ Raman Laser Generating Average Output Power of 3.14 W at 590 nm. *Optics Letters*, Vol. 32, No. 20, Oct 2007, pp. 2951-2953.

Liu, X., Zhang, H., Guo, Y. & Li, Y. (1002). Optimal Design and Applications for Quasi-Phase-Matching Three-Wave Mixing. *IEEE Journal of Quantum Electronics*, Vol. 38, No. 9, Sept 2002, pp. 1225.

Lukin, M.D. (2003). Colloquium: Trapping and manipulating photon states in atomic ensembles. Reviews of Modern Physics, Vol. 75, No. 2, April 2003, pp. 457-472.

Lukin, M.D. (2003). Colloquium: Trapping and manipulating photon states in atomic ensembles. Review of Modern Physics, Vol. 75, No. 2,April 2003, pp. 457-472.

Maker, P.D. & Terhune, R. W. (1965). Study of Optical Effects Due to an Induced Polarization Third Order In Electric Field Strength. Physical Review, Vol. 137, No. 3A, Feb 1965, pp. A801-A818.

Maneuf, S. & Reynaud, F. (1988) Quasi-Steady State Self-Trapping of First, Second and Third Order Subnanosecond Soliton Beams, *Optics Communications*, Vol. 66, No. 5,6, May 1988, pp. 325-328.

Marder, S.R., Kippelen, B., Jen, A.K.Y. & Peyghambarian, N. (1997). Design and Synthesis of Chromophores and Polymers for Electro-optic and Photorefractive Applications. *Nature*, Vol. 388, No. 6645, Aug 1997,pp. 845-851.

Martorell, J., Vilaseca, R. & Corbalan, R. (1997) Second harmonic generation in a photonic crystal. *Applied Physics Letters*, Vol. 70, No. 6, Feb 1997, pp. 702-704.

McCamant, D.W., Kukura, P., Yoon, S. & Mathies, R.A. (2004). Femtosecond Broadband Stimulated Raman Spectroscopy: Apparatus and Methods. *Review Scientific Instruments*, Vol. 75, No. 11, Nov 2004, pp. 4971-4980.

McAuslan D. L.; Ledingham P. M.; Naylor W. R.; Beavan, SE; Hedges, MP., Sellars, MJ & Longdell, J.J. (2011) Photon-echo quantum memories in inhomogeneously broadened two-level atoms, *Physical Review A*, Vol. 84, No. 2, Aug 2011, pp. 022309.

Miller, D.A.B. (2000). Optical Interconnects to Silicon. *IEEE Journal of Selected Topics in Quantum Electronics*, Vol. 6, No. 6, Nov-Dec 2000, pp. 1312-1317.

Mitchell, S.A. (2009). Indole Adsorption to a Lipid Monolayer Studied by Optical Second Harmonic Generation, *Journal of Physical Chemistry B*, Vol. 113, No. 31, Aug 2009, pp. 10693-10707.

Mondia, J.P., van Driel, H.M., Jiang, W., Cowan, A.R., & Young, J.F. (2003). Enhanced second-harmonic generation from planar photonic crystals, *Optics Letters*, Vol. 28 No. 24, Dec 2003, pp. 2500-2502.

Mormile P., Petti L., Abbate M., Musto, P., Ragosta, G. & Villano, P. (1998). Temperature switch and thermally induced optical bistability in a PDLC, *Optics Communications*, Vol. 147, No. 4-6, Feb 1998, pp. 269-273.

Mourou, G.A., Tajima, T. & Bulanov, S.V. (2006). Optics in the relativistic regime. *Review Modern Physics*, Vol. 78, April-June 2006, pp. 309–371.

Mueller, C.T. & Garmire, E. (1984). Photorefractive effect in LiNbO$_3$ Directional Couplers. *Applied Optics*, Vol. 23, No. 23, Dec 1984, pp. 4348-4351.

Nolte, D.D. (1999). Semi-insulating semiconductor heterostructures: Optoelectronic Properties and Applications, Journal of Applied Physics, Vol. 85, No. 9, May 1999, pp. 6259-6289.

Okada, A., Ishii, K., Mito, K. & Sasaki, K. (1992). Phase-Matched 2nd-Harmonic Generation in Novel Corona Poled Glass Wave-Guides. *Applied Physics Letters*, Vol. 60, No. 23, Jun 1992, pp. 2853-2855.

Ostroverkhova, O., & Moerner, W.E. (2004) "Organic photorefractives: mechanisms, materials, and applications," (2004). *Chemical Reviews*, Vol. 104, No.7, July 2007, pp. 3267–3314.

Parameswaran, K.R., Kurz, J.R., Roussev, R.V. & Fejer, M.M.(2002). Observation of 99% pump depletion in single-pass second-harmonic generation in a periodically poled lithium niobate waveguide, *Optics Letters*, Vol. 27, No. 1, Jan 2002, pp. 43-45.

Parisa, A. & Nosrat, G. (2009) All-optical ultracompact photonic crystal AND gate based on nonlinear ring resonators, *Journal of the Optical Society of America B-Optical Physics* Vol. 26, No. 1, Jan 2009, pp. 10-16.

Partovi, A. & Garmire, E.M. (1991). Band-Edge Photorefractivity in Semiconductors - Theory and Experiment. Journal of Applied Physics, Vol. 69, No. 10, May 1991, pp. 6885-6898.

Partovi, A., Garmire, E.M. & Cheng, L.J. (1987). Enhanced Beam Coupling Modulation Using the Polarization Properties of Photorefractive GaAs. *Applied Physics Letters*, Vol. 51, No. 5, Aug 1987, pp. 299-301.

Partovi, A., Millerd, J., Garmire, EM., Ziari, M., Steier, W.H., Trivedi, S.B. & Klein, M.B. (1990). Photorefractivity At 1.5-um in CdTe-V. Applied Physics Letters, Vol. 57, No. 9, Aug 1990, pp. 846-848.

Paschotta, R.(2008) Phase Matching, Encyclopedia for Photoniccs and Laser Technology. http://www.rp-photonics.com/phase_matching.html

Pask, H.M. (2003). The Design and Operation of Solid-State Raman Lasers. *Progress in Quantum Electronics*, Vol. 27, No. 1, 2003, pp. 3-56.

Penner, T.L., Motschmann, H.R., Armstrong N.J., Ezenyilimba, M.C. & Williams, D.J. (1994). Efficient Phase-Matched 2nd-Harmonic Generation Of Blue-Light In An Organic Wave-Guide. *Nature*, Vol. 367, No. 6458, Jan 1994, pp. 49-51.

Pham, A., Qiao, H., Guan, B., Gal, M., Gooding, J.J. &. Reece, P.J. (2011). Optical bistability in mesoporous silicon microcavity resonators, *Journal of Applied Physics* Vol. 109, May 2011, pp. 093113

Preston, S.G., Sanpera, A., Zepf, M., Blyth, W.J., Smith, C.G., Wark, J.S., Key, M.H., Burnett, K., Nakai, M., Neely, D. & Offenberger, A.A. (1996). High-Order Harmonics of 248.6-nm KrF Laser from Helium and Neon Ions. *Physical Review A*, Vol. 53, No. 1, Jan 1996, pp. R31-R34.

Ranka, J.K., Windeler, R.S. & Stentz, A.J. (2000). Visible continuum generation in air-silica microstructure optical fibers with anomalous dispersion at 800 nm. *Optics Letters*, Vol. 25, No. 1, Jan 2000, pp. 25-27.

Scala, M.B., Militello, B., Messina, A. & Vitanov, N.V. (2011). Stimulated Raman adiabatic passage in a system in the presence of quantum noise, *Physical Review A* Vol. 83, Jan 2011, 012101, 1-8.

Segev, M., Valley, G.C., Crosignani, B., DePorto, P. & Yariv, A. (1994). Steady-State Spatial Screening Solitons in Photorefractive Materials with External Applied-Field. *Physical Review Letters* Vol. 73, No. 24, Dec 1994, pp. 3211-3214.

Service, R.F. (2005). New Generation of Minute Lasers Steps into the Light. *Science*, Vol. 307, No. 5715, March 2005, pp. 1551-1552.

Sheik-bahae, M., Said, A.A., Wei, T-H., Hagand, D.J., Van. Stryland, E.W. (1990). Sensitive Measurement of Optical Nonlinearities Using a Single Beam *IEEE Journal of Quantum Electronics*, Vol. 26, No. 4, Apr 1990, pp. 760-769.

Shen, Y.R. (1994). Surfaces Probed By Nonlinear Optics. *Surface Science*, Vol. 299, No. 1-3, Jan 1994, pp. 551-562.

Shen, Y.R. & Bloembergen, N. (1965). Theory of Stimulated Brillouin and Raman Scattering. *Physical Review*, Vol. 137, No. 6A, March 1965, pp. A1787-A1805.

Shiraki, K., Ohashi, M. & Tateda, M. (1996). SBS Threshold of a Fiber with a Brillouin Frequency Shift Distribution. *Journal of Lightwave Technology*, Vol. 14, No. 1, Jan 1996, pp. 50–57.

Short, R.W. & Epperlein, E.M. (1992). Thermal Stimulated Brillouin Scattering in Laser-Produced Plasmas. *Physical Review Letters*, Vol. 68, No. 22, June 1992, 3307-3310.

Shu, J., Joyce, L., Fleischer J.W., Siviloglou, G.A. & Christodoulides, D.N. (2010). Diffusion-Trapped Airy Beams in Photorefractive Media, *Physical Review Letters*, Vol. 104, No. 25, Jun 2010, pp. 253904,1-4.

Smith, D. (1977). High-Power Laser Propagation - Thermal Blooming. *Proceedings of the IEEE*, Vol. 65, No. 12, Dec 1977, pp. 1679-1714.

Sohler, W. & Suche, H. (1978). 2nd-Harmonic Generation In Ti-Diffused LiNbO$_3$ Optical-Waveguides With 25-Percent Conversion Efficiency. *Applied Physics Letters*, Vol. 33, No. 6, Aug 1978, pp. 518-520.

Song, X.H., Yang, W.F.. Zeng, Z.N., Li, R.X. & Xu, Z.Z. (2010). Unipolar half-cycle pulse generation in asymmetrical media with a periodic subwavelength structure *Physical Review A*, Vol. 82, No. 5, Nov 2010, pp. 053821, 1-5.

Stegeman, G.I. , Wright, E.M. , Seaton, C.T. , Moloney, J.V. , Shen, T.P. , Maradudin, A.A. & Wallis, R. F. (1986). Nonlinear Slab-Guided Waves in Non-Kerr-Like Media. *IEEE Journal of Quantum Electronics*, QE-22, No. 6, June 1986, pp. 977-983.

Stegeman, G.I. & Segev, M. (1999). Optical Spatial Solitons and their Interactions: Universality and Diversity. *Science*, Vol. 286, No. 5444, Nov 1999, pp. 1518-1523.

Stegeman, G.I., Hagan, D.J. & Torner, L. (1996). Chi((2)) Cascading Phenomena and Their Applications to All-Optical Signal Processing, Mode-Locking, Pulse Compression and Solitons. *Optical and Quantum Electronics*, Vol. 28, No. 12, Dec 1996, pp. 1691-1740.

Steinhausser, B., Brignon, A., Lallier, E., Huignard, J.P. & Georges, P. (2007). High Energy, Single-Mode, Narrow-Linewidth Fiber Laser Source using Stimulated Brillouin Scattering Beam Cleanup. *Optics Express*, Vol. 15, No. 10, May 2007, pp. 6464-6469.

Taya, M., Bashaw, M., Fejer, M.M., Segev, M. & Valley, G.C. (1995). Observation of Dark Photovoltaic Spatial Solitons. *Phys, Review A*, Vol. 52, No. 4, Oct 1995, 3095-3100.

Tien, P.K., Ulrich, R. & Martin, R.J. (1970). Optical Second Harmonic Generation in Form of Coherent Cerenkov Radiation from a Thin-Film Waveguide. *Applied Physics Letters*, Vol. 17, No. 10, Nov 1970, pp. 447-449.

Torres, J., Coquillat, D., Legros, R., Lascaray, J.P., Teppe, F., Scalbert, D., Peyrade, D., Chen, Y., Briot, O., d'Yerville, M.L., Centeno, E., Cassagne, Cassagne, D. & Albert, J.P. (2004). Giant second-harmonic generation in a one-dimensional GaN photonic crystal. *Physical Review B*, Vol. 69, No. 8, Feb 2004, pp. 085105(1-5).

Torruellas, W.E., Wang, Z., Hagan, D.J., VanStryland, E.W., Stegeman, G.I., Tornner, L. & Menyuk, C.R. (1995). Observation of Two-dimensional Spatial Solitary Waves in a Quadratic Medium. *Physical Review Letters*, Vol. 74, No. 25, June 1995, pp. 5039.

Troccoli, M., Belyanin, A., Capasso, F., Cubukcu, E., Sivco, D.L. & Cho, A.Y. (2005). Raman Injection Laser. *Nature*, Vol. 433, No. 7028, Feb 2005, pp. 845-848.

Tutt, L.W. & Boggess, T.F. (1993). A Review of Optical Limiting Mechanisms and Devices Using Organics, Fullerenes, Semiconductors and Other Materials. *Progress in Quantum Electronics*, Vol. 17, No. 4, 1993, pp. 299-338.

Wang, G.Y., Yang, Y. & Garmire, E. (1995). Experimental-Observation Of Efficient Generation Of Femtosecond 2nd-Harmonic Pulses. *Applied Physics Letters*, Vol. 66, No. 25, Jun 1995, pp. 3416-3418.

Wang, R.Z., Guo, K.X., Liu, Z.L., Chen, B. & Zheng, Y.B., (2009). Nonlinear optical rectification in asymmetric coupled quantum wells, *Physics Letters A*, Vol. 373, No. 7, Feb 2009, pp. 795-798.

Wright, E.M., Lawrence, B.L., Torruellas, W. & Stegeman, G. (1995). Stable self-trapping and ring formation in polydiacetylene para-toluene sulfonate. *Optics Letters*, Vol. 20, No. 24, December 15, 1995, pp. 2481-2483.

Xu, C., Zipfel, W., Shear, J.B., Shear, J.B., Williams, R.M. & Webb, W.W. (1996). Multiphoton fluorescence excitation: New spectral windows for biological nonlinear microscopy. *Proceedings of the National Academy of Sciences*, Vol. 93, No. 20, Oct 1996, pp. 10763-10768.

Yang, C.M., Mahgerefteh, D., Garmire, E., Chen, L., Hu, K.Z. & Madhukar, A. (1994). Sweep-Out Times of Electrons and Holes in an InGaAs/GaAs Multiple-Quantum-Well Modulator. *Applied Physics Letters*, Vol. 65, No. 8, Aug 1994, pp. 995-997.

Yanik M.F. & Fan, S. (2003), High contrast all optical bistable switching in photonic crystal micro cavities, *Appl. Phys. Lett.*, vol. 83, no. 14, Jul 2003, pp. 2739–2741.

Yariv, A. (1978). Phase Conjugate Optics and Real-Time Holography. *IEEE Journal of Quantum Electronics*, Vol. 14, No. 9, Sept 1978, pp. 650-660.

Yoo, S.J.B. (1996). Wavelength conversion technologies for WDM network applications. *Journal of Lightwave Technology*, Vol. 14, No. 6, June 1996, pp. 955-966.

Yu, S., Zhang, Y.J., Zhang, H. & Gu, W. (2007). A tunable wavelength-interchanging cross-connect scheme utilizing two periodically poled LiNbO3 waveguides with double-pass configuration, *Optics Communications*, Vol. 272, No. 2, Apr 2007, pp. 480-483.

Zewail, A.H. (1980). Optical Molecular Dephasing - Principles of and Probings by Coherent Laser Spectroscopy. *Accounts of Chemical Research*, Vol. 13, No. 10, Oct 1980, pp. 360-368.

Zhang, Y., Gao, Z.D., Qi, Z., Zhu, S.N. & Ming N.B. (2008). Nonlinear Cerenkov Radiation in Nonlinear Photonic Crystal Waveguides, *Physical Review Letters*, Vol. 100, Apr 2008, pp. 163904,1-4.

Zhang, L., Xu, S.G., Yang, Z. & Cao, S.K. (2011) Photorefractive effect in triphenylamine-based monolithic molecular glasses with low T(g), Materials Chemistry and Physics, Vol. 126 No. 3 Apr 2011, pp. 804-810.

Zipfel, W.R., Williams, R.M., Christie, R., Nikitin, A.Y., Hyman, B.T., & Webb, W.W. (2003). Live tissue intrinsic emission microscopy using multiphoton-excitenative fluorescence and second harmonic generation. *Proceedings of the National Academy of Sciences of The United States of America*, Vol. 100, No. 12, Jun 2003, pp. 7075-7080.

2

Anisotropic Second- and Third-Order Nonlinear Optical Response from Anisotropy-Controlled Metallic Nanocomposites

Roberto-Carlos Fernández-Hernández[1], Lis Tamayo-Rivera[1],
Israel Rocha-Mendoza[2], Raúl Rangel-Rojo[2],
Alicia Oliver[1] and Jorge-Alejandro Reyes-Esqueda[1]
*[1]Institute of Physics, National Autonomous University of Mexico,
Circuit for Scientific Research S/N, University City, Mexico City,
[2]Department of Optics, Center for Scientific Research and Education
Superior de Ensenada, Ensenada,
Mexico*

1. Introduction

Plasmonics has become a route to develop ultracompact optical devices on a chip by using extreme light concentration. It also gives the ability to perform simultaneous electrical and optical functions, and facilitates dramatic enhancement of localized field intensities via metallic nanostructures. On the other hand, nonlinear optical interactions scale with the local intensity of the optical field, i.e., the dielectric polarization of a given material responds to the local electric field in a high-order way. For a metal-dielectric nanostructure, the combination of plasmonics with its nonlinear optical response offers the opportunity to manipulate nonlinear optical responses in sub-diffraction-limited volumes.

Nevertheless, before taking advantage of such opportunity, a strong understanding of the nonlinear optical response due to metallic nanoparticles is necessary (Fernández-Hernández et al., 2011; Rangel-Rojo et al., 2009, 2010; Reyes-Esqueda et al., 2009; Rodríguez-Iglesias et al., 2009; Torres-Torres et al., 2008). On doing so, the possibility of conferring anisotropic symmetries to metallic nanocomposites (Oliver et al., 2006; Rodríguez-Iglesias et al., 2008; Silva-Pereyra et al., 2010) revealed also the anisotropy of their linear and nonlinear optical responses (Fernández-Hernández et al., 2011; Rangel-Rojo et al., 2009, 2010; Reyes-Esqueda et al., 2008, 2009; Rodríguez-Iglesias et al., 2009). In particular, the anisotropic nonlinear optical response, for given wavelength and incident polarization, reveals a complex contribution from all the different, nonzero, linearly independent third-order susceptibility tensor's components (Reyes-Esqueda et al., 2009; Rodríguez-Iglesias et al., 2009). Besides, given this anisotropy, although the nanocomposite remains being centrosymmetric, there is the possibility of measuring a nonzero optical second-order nonlinearity in the form of a second harmonic generation (SHG) signal, as it has been shown in previous results (Aktsipetrov et al., 1995; Brevet et al., 2011; Dadap et al., 1999, 2004; Figliozze et al., 2005; Gallet et al., 2003; Mendoza et al., 2006), and also verified by us experimentally (Rocha-Mendoza et al., 2011).

Therefore, in this chapter we first present an overview of recent studies about the nonlinear optical response of metallic nanoparticles. Then, we present recent previous results regarding the third-order nonlinear optical response of metallic nanocomposites. After that, we discuss the dependence of this response on the tensor's components, making emphasis on the anisotropic case. Finally, we calculate the general form of the second-order susceptibility tensor for anisotropic nanocomposites, together with some experimental results for the polarization dependence of second harmonic generation in these composites.

2. Nonlinear optical response of metallic nanoparticles

In recent years, nanostructured materials composed of metal nanoparticles (NPs) have attracted much attention due to the possibility of using their nonlinear optical properties for photonic nanodevices (Inouye et al., 2000; Matsui, 2005) and plasmonic circuitry (Barnes et al., 2003; Tominaga et al., 2001). Their linear and nonlinear optical properties are dominated by collective electron-plasma oscillations, the so-called localized surface plasmon resonances (LSPRs), and a vast literature can be found elsewhere studying such properties (Aktsipetrov et al., 1995; Barnes et al., 2003; Brevet, 2011; Dadap et al., 1999; Inouye et al., 2000; Karthikeyan et al., 2008; Kim et al., 2006; Matsui, 2005; McMahon et al., 2007; Rangel-Rojo et al., 2009, 2010; Ryasnyansky et al., 2006; Tominaga et al., 2001; Zheludev & Emelyanov, 2004). In particular, in 2008, when studying Cu NPs embedded in a silica matrix using nanosecond and picosecond light pulses, we found that thermal effects for the nanosecond regime, and induced polarization for the picosecond one, were the physical mechanisms responsible for the saturable optical absorption and the Kerr effect presented by the nanocomposites (Torres-Torres et al., 2008). However, by following Hache, et al. (Hache et al., 2004), we also remarked the contribution of the hot-electrons generation to the observation of saturable absorption. From there, we have widely studied the optical third-order nonlinearity of randomly arranged, but elongated and aligned in a preferential direction, metallic NPs embedded in silica (Fernández-Hernández et al., 2011; Rangel-Rojo et al., 2009, 2010; Reyes-Esqueda et al., 2009; Rodríguez-Iglesias et al., 2009). We have put especial attention, on one hand, to the contribution to these properties from the electron transitions charactheristic of metallic NPs, that is, intra- and inter-band transitions, but also to the contribution from the mentioned hot-electrons generation (Fernández-Hernández et al., 2011). From here, the dependence of these nonlinearities on the incident wavelength and irradiance has been observed, allowing also the observation of sign switching of the nonlinear absorption and refraction (Fernández-Hernández et al., 2011). On the other hand, the shape-anisotropy of the ion-deformed metallic NPs embedded in a dielectric matrix (Oliver et al., 2006; Rodríguez-Iglesias et al., 2008; Silva-Pereyra et al., 2010) induces also an anisotropic third-order nonlinear optical response, as we have shown for different temporal regimes, in the case of Ag and Au (Fernández-Hernández et al., 2011; Rangel-Rojo et al., 2009; Reyes-Esqueda et al., 2009; Rodríguez-Iglesias et al., 2009). For the anisotropic nanocomposites the analysis is now more complex: there are two LSPRs, one associated to the major axis of the elongated NP, and another one associated to its minor axis. This fact duplicates the contribution from the intra-band transitions, but also that from the hot-electrons generation. Besides, the form of the nonlinear susceptibility tensor becomes very specific, having now only three linearly independent, nonzero components (Reyes-Esqueda et al., 2009; Rodríguez-Iglesias et al., 2009), where each one of them may be measured depending on the incident polarization of the exciting beam, and on the angular position of the nanocomposite with respect to the beam's wavevector. We have shown also the possibility of understanding the nonlinear optical absorption behavior by using a two-level model (Rangel-Rojo et al., 2009). Thus, for the third-order nonlinear optical response from isotropic and anisotropic metallic NPs, besides of being very large, the sign of both the nonlinear absorption and refraction can change, depending on the wavelength, irradiance and incident polarization (Fernández-Hernández et al., 2011).

Regarding the second-order nonlinear optical response, metallic NPs are centrosymmetrical materials, their crystalline lattice structures are cubic face-centered and, in principle, no SHG from the bulk NP takes place in the electric dipole approximation. The SHG origin of such materials is attributed then to higher order interactions like electric quadrupole and magnetic dipole responses from the NPs bulk and/or electric dipole responses allowed from the NPs surfaces (Aktsipetrov et al., 1995; Brevet et al., 2011; Dadap et al., 1999, 2004; Figliozze et al., 2005; Gallet et al., 2003; Mendoza et al., 2006), where the inversion symmetry of the bulk material is broken. The latter response dominates in the specific case when the NP size is much smaller than the wavelength of the exciting (fundamental) beam (Aktsipetrov et al., 1995) so that field retardation effects (no spatial dependence of the electromagnetic fields) are neglected. Therefore, the problem at the macroscopic level turns to be very similar to that of nonlinear media containing particles of non-centrosymmetrical material apart from the interfacial second-order origin of the response. In a sense, the arrangement of the NPs in the array resembles that of the atoms in a crystal cell, where phase-matched SHG signal radiates in specific directions. This principle has been utilized for example on planar structures containing metallic 2D arrays of nanoparticles lacking inversion symmetry, providing coherent addition of the SH field where the efficiency of the process increased rapidly with decreasing nanoparticle size (McMahon et al., 2007; Zheludev et al., 2004). In this context, we demonstrate that, similar to the analysis to derive molecular orientation information of smaller noncentrosymmetic units at interfacial monolayer's or macromolecules systems using SHG/Sum-Frequency Generation (SFG) experiments (Knoesen et al., 2004; Leray et al., 2004; Psilodimitrakopoulos et al., 2009; Campagnola & Loew, 2003; Rocha-Mendoza et al., 2007; Shen, 1989; Zhuang et al., 1999), we can treat the actual submicrometric layer containing randomly arranged, but highly aligned anisotropically shaped Ag-NPs (elongated or nearly spherical), as a nonlinear media; where the origin of its macromolecular second-order susceptibility, $\chi^{(2)}$, is the coherent contribution of the SH signal induced on every single nanoparticle.

3. Third-order nonlinear optical response from isotropic metallic nanocomposites

In order to get more control of the nonlinear optical response from metallic nanocomposites, very recently we carefully studied the wavelength dependence of this response (Fernández-Hernández et al., 2011). To do that, we performed closed- and open-aperture Z-scan measurements at several wavelengths in isotropic and anisotropic metallic nanocomposites produced by ion implantation. We have observed dramatic changes of sign for both nonlinear refraction and absorption, when passing from Au to Ag and/or varying the wavelength and irradiance. The results put into evidence the hot-electron contribution to the nonlinear optical response, when compared to inter- and intra-band transitions contribution. In this section, we present the main results obtained concerning the isotropic nanocomposites, leaving those from the anisotropic ones for the next section.

3.1 Synthesis of isotropic metallic nanocomposites and optical measurements

As reported before (Oliver et al., 2006; Reyes-Esqueda et al., 2009), high-purity silica glass plates where implanted at 0° of incidence at room temperature with 2 MeV Ag^{2++} (or Au^{2++}) ions at a fluence of 3.35×10^{16} ion/cm² (3.10×10^{16} ion/cm²). The depth of the Ag NPs layer was 0.94 µm with a FWHM of 0.72 µm, while for Au the depth was 0.57 µm with a FWHM of 0.36 µm. After Ag implantation, the samples were thermally annealed for 1 hr in a reducing atmosphere $50\%H_2+50\%N_2$ at a temperature of 600°C for Ag. In the case of Au, an oxidizing atmosphere (air) was used for 1 hr at 1100°C. The metal implanted distributions and fluences were determined by Rutherford Backscattering Spectrometry (RBS) measurements using a 3 MeV $^4He^+$ beam for Ag and a 2 MeV $^4He^+$ beam for Au. Ion implantation and RBS analysis were performed using the IFUNAM's 3MV Tandem accelerator NEC 9SDH-2 Pelletron facility.

Linear optical absorption measurements were performed with an Ocean Optics Dual Channel SD2000 UV-visible spectrophotometer at normal incidence on the surface sample, changing only the incident polarization. The third-order nonlinear optical spectroscopy was performed by the Z-scan method (Sheik-Bahae et al., 1990) at 355, 500, 532, 600 and 750 nm. A picosecond pulsed laser system (PL2143A, 26ps) and an optical parametric generator (PG 401/SH), both from EKSPLA, were used as light sources. These sources were focused with a focal length of 500 mm, where in each case the beam waist was measured by means of the knife's edge method (Khosrofian & Garetz, 1983). The Rayleigh length was calculated to be around 1 cm for all the wavelengths considered, much larger than the thickness of the samples. It was verified that the nonlinear optical response from the SiO_2 matrix was negligible when compared to that from the nanocomposite (matrix + NPs layer). The reference and transmitted beams (open- and closed-aperture) were measured with Thorlabs DET210 fast photodiodes. All optical measurements were performed in the Nonlinear Optics Laboratory at IFUNAM.

3.2 Main results

The linear absorption spectra of the isotropic nanocomposites are shown in Fig. 1. The invariance of the optical absorption for different incident polarizations confirms the optical isotropic behavior. In these Fig., it is also shown the wavelength position of the laser beam used for performing the Z-scan measurements in each case.

Since the plasmonic systems usually exhibit positive and negative nonlinear absorption (NLA) at the same time, a fact that clearly influences the closed-aperture Z-scan traces for nonlinear refraction (NLR), we determined the nonlinear optical coefficients by a theoretical fitting, instead of using the approximate relationships that appear in reference (Sheik-Bahae et al., 1990), as it is usually done. For NLA, the fitting was made following (Liu et al., 2001;

Wang et al., 2010), where the presence of saturable and induced absorption is modeled through the following relationship:

$$\frac{dI}{dz'} = -\left[\frac{\alpha_0}{(1+I/I_0)} + \beta I\right]I,$$ (1)

where α_0 is the linear absorption coefficient at the proper wavelength, β is the two-photon absorption coefficient, I is the intensity of the laser, and I_s is the saturation intensity for the saturation of absorption process, respectively. Normally, in the case of third order nonlinear optical phenomena, the NLR contribution is written as $n=n_0+n_2I$, and the NLA as $a=a_0+\beta I$, where n_0 and a_0 are the linear refraction and absorption, respectively; n_2 is the nonlinear refractive index and β is the two-photon absorption coefficient. But when we have saturation of absorption, the first part of Eq. 1 can be developed in series in order to obtain $a\approx a_0 (1-I/I_s)$, where we define a negative NLA coefficient as $\beta=-a_0/I_s$. Then, we can note that, for a proper irradiance, the NLA superposition can result in cancellation of any nonlinear absorption effects. In order to show the nonlinear optical response through the entire spectral range studied, we resume the results in Figs. 2 and 3, where we have included the linear absorption spectra in the graphics in order to see in which zone we were exciting, near or far of the plasmon resonance. It is worth mentioning that we have shown in Fig. 3, the negative and positive values of β obtained when fitting, by using Eq. 1, the experimental data obtained from the z-scan measurements. This will help us later when we discuss the physical origin of the observed nonlinear optical behavior for both plasmonic nanocomposites.

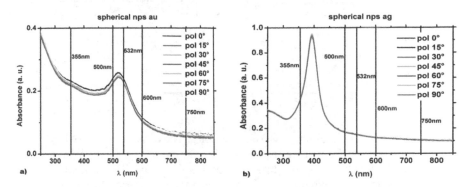

Fig. 1. Linear optical absorption of isotropic samples. a) Au NPs. b) Ag NPs.

In Fig. 2, for the Au nanocomposite, we can see that the NLR is always positive and grows with wavelength, except at the shortest wavelength used (355 nm), where the sign changes. In the case of Ag, the NLR is also positive, but there is a change of sign at 500 nm, near the low-energy side of the plasmon resonance. This change of sign was also found at 532 nm for low irradiances, but it became positive when increasing the irradiance. Then, we argue that, for a higher irradiance at 500 nm, the sign may become positive. This has to do with the type of electronic transition being excited in each case, as it will be discussed later. For NLA (figure 3), a superposition of effects for both types of nanocomposite is clearly seen, confirming that, with the proper irradiance, the nonlinear absorption can be cancelled, as shown for Ag at 600 nm and for Au at 355 nm. It can be also deduced that, for low irradiances, a positive NLA dominates, whereas at higher irradiances, NLA changes sign, *i.e.*, the nanocomposite's absorption saturates and becomes more transparent.

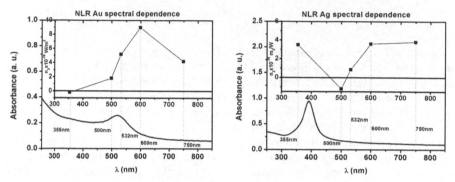

Fig. 2. NLR spectroscopy results for isotropic samples. The upper data of each part show the value of n_2 obtained from the measurements for Au and Ag NPs isotropic samples, while the lower curve is the corresponding linear optical absorption. The dotted vertical lines represent the wavelength where the NLR measurements were performed.

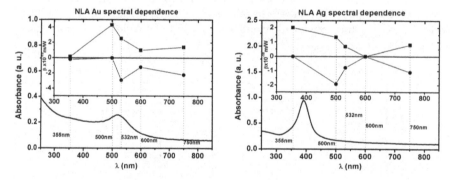

Fig. 3. NLA spectroscopy results for isotropic samples. The upper data of each part show the value of β obtained from the measurements for Au and Ag NPs isotropic samples, while the lower curve is the corresponding linear optical absorption. The dotted vertical lines represent the wavelength where the NLA measurements were performed.

3.3 Discussion

It is well known that the optical response of metallic nanocomposites depends on the excitation wavelength, since the induced effects in the NPs can excite inter- and/or intra-band electronic transitions. In the linear regime, it has been well established that plasmon resonances are due to intra-band transitions, and that they can be explained using Drude's model for metals, by adding a correction of the inter-band transitions when calculating the dielectric function of the NPs (Noguez, 2007). Regarding the nonlinear regime, in the isotropic case, both electronic transitions contribute with the same sign to the nonlinear optical response near the plasmon resonance, as explained by Hache *et al.* (Hache et al.,

1986, 1988). Nevertheless, intra-band transitions are present in the whole visible spectrum, while inter-band only when the respective band-gap is overcame. In the case of gold, this band gap corresponds to 1.7 eV (729 nm) (Hache et al., 1988).

Additionally, there is another contribution to the optical properties in the nonlinear regime. This is the formation of hot-electrons, which contribute with opposite sign to the optical response. Following Hache *et al.* (Hache et al., 1988), near the plasmon resonance, the nonlinear optical response is due, principally, to inter-band transitions and the formation of hot-electrons. Then, both contributions were present when we use 532 nm for Au and 355 nm for Ag, since these wavelengths lay near of the respective plasmon resonances. At low irradiances dominates the inter-band contribution, which is reflected in a positive NLA, but when the irradiance is increased, the formation of hot-electrons increases at the same rate, dominating the optical response and changing the sign of the NLA.

For longer wavelengths, the nonlinear optical response is very similar for both nanocomposites, due to the absence of inter-band transitions. In that region, only the free-electron response, explained by the Drude model, contributes to the optical response, *i.e.*, the intra-band transitions and the hot-electrons. But, because of the low energy of the photons in this region, the hot-electron contribution is not as strong as near the plasmon resonance and, consequently, a positive NLA is always dominant. However, we must consider that the shorter the wavelength, the higher the energy of the photon, consequently, the production of hot-electrons is higher for shorter wavelengths, and the positive NLA ceases to be dominant for a given irradiance and different wavelengths (Fernández-Hernández et al., 2011). For NLR, there is a simple calculation of the NLR index following Boyd (Boyd, 2008), which is given by

$$n_2 = \frac{-e\alpha_0\tau_r}{2\varepsilon_0 n_0 mh\omega^3},$$

(2)

where e is the charge of the electron and τ_r is the response time of the material (300 fs). Introducing the respective values for our nanocomposites, this shows that n_2 is around 10^{-16} m^2/W, of the order of magnitude found in the measurements, as shown in Fig. 2. The entire nonlinear optical response discussed here is due completely to the presence of the NPs into the matrix, because, as it was said before, the nonlinear optical response of the matrix was negligible.

4. Third-order nonlinear optical response from anisotropic metallic nanocomposites

The anisotropy of the metallic nanocomposites makes their nonlinear optical response more complex. Now the dependence on the angular position of the sample and on the incident polarization, due to the form of the third-order nonlinear susceptibility tensor, may cause that the contributions to this response come not only from the inter-band transitions, but also from the intra-band transitions and the hot-electron contribution from two surface plasmon resonances, the one associated to the minor axis and the one associated to the major one.

4.1 Synthesis of anisotropic metallic nanocomposites and optical measurements

For the anisotropic nanocomposites, the previously implanted silica plates were cut into several pieces and each piece was irradiated at room temperature with 8 MeV Si ions at a fluence of 5.0×10^{15} ions/cm^2 for Ag, and 10 MeV Si ions at a fluence of 1.2×10^{16} ions/cm^2 for Au. The Si irradiation was performed under an angle off normal of $\theta = (0.0°$ or $80.0° \pm 0.5°)$ for both, Ag and Au. As it has shown before in previous works (Oliver et al., 2006; Rodríguez-Iglesias et al., 2008), this irradiation deforms the formed NPs, transforming them in prolate spheroids all oriented along the direction of irradiation. The Si irradiation was also performed using the IFUNAM's 3MV Tandem accelerator NEC 9SDH-2 Pelletron facility. Regarding the nonlinear optical measurements, they were performed for different incident polarizations, 0°, 45° and 90° for normal incidence of light, where 0° corresponds to the polarization aligned with major axis of the NP, and 90° to polarization aligned with the NP minor axis, for deformation at 80°. This convention was also maintained for nanocomposites deformed at 0°.

4.2 Results

The linear absorption spectra of the anisotropic nanocomposites, where the NPs were deformed at 0° and 80°, respectively, are shown in Figs. 4 and 5. The spectra for different polarizations show isotropy for deformation at 0°, as it was expected, because the light's

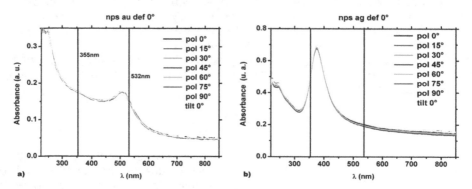

Fig. 4. Linear optical absorption of anisotropic samples with NPs deformed at 0°.
a) Au NPs. b) Ag NPs.

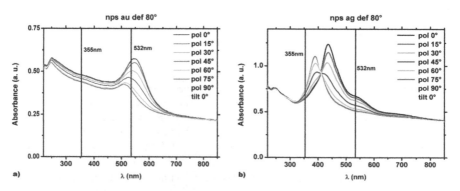

Fig. 5. Linear optical absorption of anisotropic samples with NPs deformed at 80°.
a) Au NPs. b) Ag NPs.

polarization is parallel in all cases to the minor axis of the deformed NPs; while the anisotropic optical behavior is clear for deformation at 80° when varying the polarization. This last behavior can be explained with the presence of a second plasmon resonance at higher wavelengths, which corresponds to the major axis of the NPs. As before, in these Figs., it is also shown the wavelength position of the laser beam used for performing the Z-scan measurements in each case.

Regarding the nonlinear response of the anisotropic plasmonic nanocomposites, Figs. 6-9 show the results for both type of nanocomposites, deformed at 0° and 80°, where we have used only the wavelengths of 532 nm and 355 nm, as indicated previously in Figs. 4-5. As said before, the incident polarization angles used were 0°, 45° and 90° for normal incidence of light, with 0° corresponding to polarization aligned with major axis of the NP, and 90° to polarization aligned with the NP minor axis, for deformation at 80°.

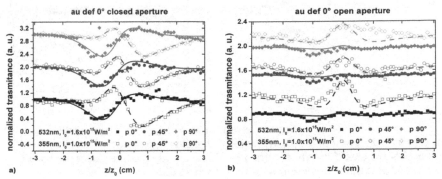

Fig. 6. a) Closed- and b) open-aperture Z-scan for the anisotropic nanocomposite with Au NPs deformed at 0°. Different symbols are used for the different incident polarizations. The empty symbols are used for data obtained at 355 nm, while the filled ones are used for data obtained at 532 nm. Lines represent the fitting to these data. The data sets for different polarizations have been vertically shifted for clarity.

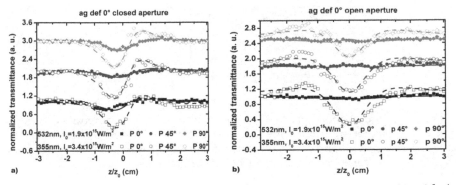

Fig. 7. a) Closed- and b) open-aperture Z-scan for the anisotropic nanocomposite with Ag NPs deformed at 0°. Different symbols are used for the different incident polarizations. The empty symbols are used for data obtained at 355 nm, while the filled ones are used for data obtained at 532 nm. Lines represent the fitting to these data. The data sets for different polarizations have been vertically shifted for clarity.

Figures 6-7 show the nonlinear spectroscopy in the case of deformation at 0°, for both Au and Ag. It is remarkable the independence of the measurements on the incident polarization, confirming the isotropy of the deformed NPs into its transversal section. In the case of Au, there is a clear change of sign in NLR when passing from 532 nm to 355 nm, as in the isotropic case, where the former wavelength lays on the plasmon resonance and the last into the inter-band electronic transitions region. In the case of NLA, a superposition of both saturable and induced absorption is clearly observed for both wavelengths; but at 532 nm, for comparable irradiances, these effects almost cancel each other in both nanocomposites.

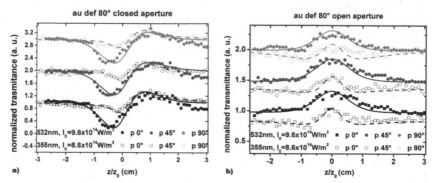

Fig. 8. a) Closed- and b) open-aperture Z-scan for anisotropic nanocomposite with Au NPs deformed at 80°. Different symbols are used for the different incident polarizations. The empty symbols are used for data obtained at 355 nm, while the filled ones are used for data obtained at 532 nm. Lines represent the fitting to these data.

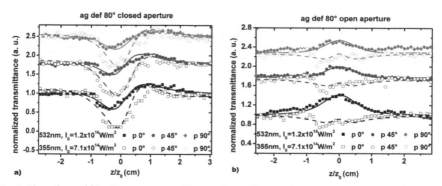

Fig. 9. a) Closed- and b) open-aperture Z-scan for anisotropic nanocomposite with Ag NPs deformed at 80°. Different symbols are used for the different incident polarizations. The empty symbols are used for data obtained at 355 nm, while the filled ones are used for data obtained at 532 nm. Lines represent the fitting to these data.

In the case of NPs deformed at 80°, from Figs. 8 and 9, it can be seen that, for NLR, in both nanocomposites, there is not a change of sign, but the magnitude of the response is larger when exciting near the plasmon resonances, 532 nm in the case of Au nanocomposite, and 355 nm in the case of Ag one. For NLA, a superposition of effects with different sign is present again in both samples, where saturable absorption dominates at 532 nm.

4.3 Analysis

Regarding the anisotropic nanocomposites, as reported recently (Reyes-Esqueda et al., 2009; Rodríguez-Iglesias et al., 2009), there are three linearly independent components of the third order nonlinear susceptibility tensor that are responsible for the optical response of the nanocomposite, one of them corresponding to the mayor axis of the NP, another to the minor axis, and a final one to a linear superposition of them. Consequently, the third-order nonlinear polarization for the anisotropic nanocomposite may be written in terms of the angle between the electric field and the NP's axis, which defines the optical axis of the nanocomposite, as

$$\mathbf{P}_{NL}^{(3)}(\theta;\omega) = 3|E(\omega)|^2 E(\omega)\left[\chi_{1111}^{(3)}\sin^3\theta\hat{\mathbf{j}} + \frac{3}{2}\chi_{1133}^{(3)}\sin 2\theta\{\cos\theta\hat{\mathbf{j}} + \sin\theta\hat{\mathbf{k}}\} + \chi_{3333}^{(3)}\cos^3\theta\hat{\mathbf{k}}\right]. \quad (3)$$

When the incident electric field is parallel to the minor axis, the nonlinear polarization is trivially given by

$$\mathbf{P}_{NL}^{(3)}(\theta;\omega) = 3|E(\omega)|^2 E(\omega)\chi_{1111}^{(3)}\hat{\mathbf{i}}. \quad (4)$$

These two equation allows us to analyze the results presented in Figs. 6-9. For the case of deformation at 0°, the incident electric field is clearly parallel to the minor axis of the NP, being then Eq. 4 the correct one. According to our previous results (Silva-Pereyra et al., 2010), the minor axis of the NP has a size very similar to the radius of the isotropic NPs. Then, in this case, independently of the polarization, the incident light sees always an isotropic metallic system. Therefore, one can explain the results presented at Figs. 6 and 7 in a similar way as explained for the isotropic case. For Au, at 355 nm, the inter-band transitions contribute the most to the nonlinear optical response. At 532 nm, the intra-band transitions and the hot-electron contribution dominate now the optical response. For Ag, at 355 nm, the intra-band transitions and the hot-electron contribution dominate now; but, for 532 nm, there is only the intra-band transition contributing to the optical response. For both cases, one can see the same behavior as for the isotropic case.

But for deformation at 80°, this analysis is more difficult. Now, for incident polarization of 0°, Eq. 3 is the correct one to try to explain the nonlinear optical response observed. In this case, since the measurements are performed at normal incidence, the angle between the electric field and the major axis of the NP is 10°. By evaluating this into Eq. 3, one can observe that the larger contribution to the observed behavior comes from the term corresponding to the major axis, that is, $\chi_{3333}^{(3)}$, although there is also a contribution from the term mixing the axes, $\chi_{1133}^{(3)}$, and then the contribution from the plasmon resonances. Fortunately, for the wavelengths used in this work, the analysis may be done considering only one plasmon resonance at a time. This fact allows making the same considerations as for the isotropic case, and, for applications, by varying wavelength, polarization and angular position of the sample, one can switch the sign of the nonlinear optical response of these nanocomposites. For example, for Ag at 355 nm, the resonance associated to the major axis almost does not contribute to the optical response since it is rather far from this wavelength. Similarly, the inter-band transitions do not contribute. Then, the observed response is mostly due to the intra-band transitions and the hot-electron contributions from the resonance associated to the minor axis. But this is exactly what happened for the isotropic case and for the NPs deformed at 0°. In the case of Au at 532 nm, the inter-band transitions do not contribute, but now, both resonances contribute to the observed response, although the wavelength is nearest to the resonance associated to the major axis, which presents also a larger absorbance, increasing the magnitude of the nonlinearity measured too. For incident polarization of 90°, Eq. 4 again describes the observed response, which is again, in general, as for the isotropic case. However, for the deformation at 80°, for Au, there is one discrepancy at 355 nm, for both polarizations the NLR turns to be positive, while for deformation at 0° and for the isotropic case, this was negative. This result deserves a more careful analysis, which will be done at some other time.

5. Second-order nonlinear optical response for anisotropic nanocomposites

Although anisotropic metallic NPs are still centrosymmetrical systems, there is the possibility of measuring SHG from them when higher order interactions, like electric quadrupole, and magnetic dipole responses from the NPs bulk, and/or electric dipole responses allowed from the NPs surfaces (Aktsipetrov et al., 1995; Brevet et al., 2011; Dadap et al., 1999, 2004; Figliozze et al., 2005; Gallet et al., 2003; Mendoza et al., 2006), where the inversion symmetry of the bulk material is broken, are present. The latter response dominates in the specific case the NP size is much smaller than the wavelength of the exciting (fundamental) beam (Aktsipetrov et al., 1995), so that field retardation effects (no spatial dependence of the electromagnetic fields) are neglected.

5.1 Theoretical analysis

In general, the second-order nonlinear polarization, $P_i^{(2)}$, induced by an incident electric field, E, for the bulk second-order susceptibility, $\chi_{ijk}^{(2)}$, is given by $P_i^{(2)} = \chi_{ijk}^{(2)} E_j E_k$. For materials composed of nonlinear optical scatterers much smaller than the wavelength of the fundamental beam, such as in the case of thin films composed of syntectic macromolecules or fibrillar proteins for instance (Knoesen et al., 2004; Leray et al., 2004; Podlipensky et al., 2003; Rocha-Mendoza et al., 2007), the origins of the bulk second-order susceptibility $\chi_{ijk}^{(2)}$ comes from the coherent summation of the molecular hyperpolarizability $\beta_{i'j'k'}$ of the smaller molecules. In a similar way, due to the fact that the elongated NPs of our samples are at least a hundred times smaller than the wavelength of the fundamental light and in the limit of weak coupling between each nonlinear NP, we can express the macroscopic susceptibility of the thick layer containing NPs as

$$\chi_{ijk}^{(2)} = N \sum \left\langle R_{ijk,i'j'k'} \right\rangle \beta_{i'j'k'} . \tag{5}$$

where N is the number density of NPs contained within the point spread function of the beam, $\beta_{i'j'k'}$ is the hyperpolarizability tensor for the NPs, and $R_{ijk,i'j'k'}$ are the elements of the Euler rotation matrix that transforms the hyperpolarizability $\beta_{i'j'k'}$ in the NPs coordinate system ($i'j'k' = \xi, \eta, \zeta$) to the laboratory coordinates ($ijk = x, y, z$). The angular averaging denoted by $\langle \ \rangle$ accounts for the angular distribution of the NPs. Fig. 10 shows a schematic representation of the Euler angles, where the z-axis is normal to the sample surface and the ζ-axis is along the NP long axis. Assuming the interface between the glass substrate and the NPs layer to be azimuthally isotropic (invariance under ψ-rotation), i. e. this interface does not contribute to the second harmonic signal, there are only three nonvanishing independent components of $\chi^{(2)}$, $\chi^{(2)}_{xxz} = \chi^{(2)}_{yyz} = \chi^{(2)}_{xzx} = \chi^{(2)}_{yzy}$, $\chi^{(2)}_{zxx} = \chi^{(2)}_{zyy}$ and $\chi^{(2)}_{zzz}$.

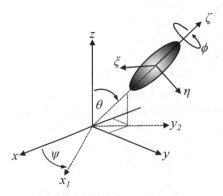

Fig. 10. Euler angles, (ψ, θ, ϕ), relating the laboratory coordinates system, xyz, and the coordinate system, $\xi\eta\zeta$, of a single El Ag-NP. ϕ is the isotropic azimuthal angle over the ζ-axis. $x1$-axis is obtained from the first rotation, ϕ. θ is the pitch angle formed from the z-axis of the laboratory system and the ζ-axis of the NP coordinate system, after the first rotation ϕ. ψ is the azimuthal angle over z-axis.

It is common practice to calculate the average orientation of the sub-molecular units that give rise to the macroscopic susceptibility of nonlinear materials using SHG/SFG experiments. In the process, analytical expressions of the independent tensor components are typically written in terms of the pitch angle, θ, defined by the z-axis and the noncentrosymmetric subunit ζ-axis, and the resulting nonzero elements of the microscopic hyperpolarizability, $\beta_{i'j'k'}$. Such expressions and their derivation are found elsewhere [16-19,24], and will not be rewritten here. Instead, for the purpose of this work, we followed the formalism used in Refs. (Hirose et al., 1992; Knoesen et al., 2004; Zhuang et al., 1999) to derive expressions for the effective susceptibility tensor, $\chi^{(2)}_{eff}$, in terms of the experimental variables α and ϕ. With α as the angle of the linearly polarized fundamental beam with respect to the plane of incidence, and, ϕ, as the angle made by the projection of the NP long axis, ζ, over the xy plane and the fixed plane of incidence contained in xz. From Fig. 10 we can deduce that $\phi = \psi + 90°$. The total SH intensity is proportional to the sum of the effective susceptibility p and s as follows

$$I_{SH} \propto \left|\chi^{(2)}_{eff}\right|^2 = \left|\chi^{(2)}_{eff,p\alpha}\right|^2 + \left|\chi^{(2)}_{eff,s\alpha}\right|^2 , \tag{6}$$

where the first and second sub indexes (from left to right) of $\chi^{(2)}_{eff}$ in the right-hand side of Eqn. 6 denote the output (fixed) and input (variable) polarization directions, respectively, and each component can be computed from

$$\chi^{(2)}_{eff,\hat{e}_2\hat{e}_1} = \left[\hat{e}_2(2\omega)\cdot\bar{L}(2\omega)\right]\cdot\chi^{(2)}:\left[\hat{e}_1(\omega)\cdot\bar{L}(\omega)\right]^2, \tag{7}$$

Here $\hat{e}_1$ and $\hat{e}_2$ are the unit polarization vectors of the fundamental and SHG beams, respectively. With $\hat{e}_1$ (or α) $=0^o$ (or 180°) for p-polarized light and 90° (or 270°) for s-polarized light, for example. $\bar{L}(\Omega)$ is the Fresnel factor at frequency Ω at the silicon substrate and nonlinear media interface. This parameter is found to be quite sensitive in the determination of molecular orientation since it depends on the index of refraction of the interfaces involved and the cosine of the angle of incidence/reflection of the fundamental/SH beam (Zhuang et al., 1999). However, we already know the orientation angle of the NPs and for simplicity, we will approximate this values to unity.

In the first approximation, we chose to model the elongated NPs as rod particles for which the hyperpolarizability tensor is cylindrically symmetric along ζ (invariance under ϕ-rotation). We also assume that the optical frequencies of the fundamental and SH beams are not in resonance with electronic transitions, so that there are only two nonvanishing independent components ($\beta_{\xi\xi\zeta} = \beta_{\eta\eta\zeta}$, and $\beta_{\zeta\zeta\zeta}$). Under these conditions using Eqs. 5 and 7 we obtain

$$\begin{aligned}
\chi^{(2)}_{eff,p\alpha} &= \frac{-1}{4}a\sin2\alpha\sin\phi\left(\cos\phi+\sin^2\phi\right)r \\
&+\frac{\sqrt{2}}{2}a\cos^2\alpha\sin^2\frac{\phi}{2}\left(r\sin^2\phi+\cos^2\phi\right) \\
&-\frac{\sqrt{2}}{2}ab\cos\phi\sin^2\phi\left(1-\frac{1}{2}\cos^2\alpha\right) \\
&+\frac{1}{4}a\sin2\alpha\sin2\phi\sin^2\frac{\phi}{2}
\end{aligned} \tag{8}$$

for p-SH and

$$\begin{aligned}
\chi^{(2)}_{eff,s\alpha} &= \frac{-\sqrt{2}}{4}a\sin^2\alpha\sin^3\phi \\
&-\frac{\sqrt{2}}{4}\sin\phi\left(1+\cos^2\phi\right)r \\
&-\frac{\sqrt{2}}{4}\sin\phi\cos^2\alpha\left(\cos\phi-\sin^2\phi\right)r \\
&+\frac{1}{2}a\sin2\alpha\left(\cos^2\phi\sin^2\frac{\phi}{2}r+\sin^2\phi\cos^2\frac{\phi}{2}\right)
\end{aligned} \tag{9}$$

for s-SH, respectively.

In Eqs. (8) and (9) $r = \beta_{\xi\xi\zeta}/\beta_{\zeta\zeta\zeta}$ and $a = N\beta_{\zeta\zeta\zeta}$ are our fitting parameters, while $b=1-r$. Note that $\chi^{(2)}_{eff,p\alpha} = \chi^{(2)}_{eff,s\alpha} = 0$, in the specific case when $\phi=0$, this result is expected since under this configuration the input electric field finds isotropically shaped NPs. For all other configurations ($\phi=90^\circ$; $\phi=180^\circ$ and $\phi=270$) $\chi^{(2)}_{eff,p\alpha} \neq \chi^{(2)}_{eff,s\alpha} \neq 0$ and polar traces with different lobes of maximum SH are found for p- and s-SH, respectively, as will be shown in our discussions.

5.2 Sample preparation and optical measurements

The anisotropic Ag nanocomposites used for SHG measurements were prepared as before, only the angle of deformation was modified to 45° and the fluences were 2.4×10^{17} ion/cm² for Ag, and 1×10^{16} ion/cm² for Si. The depth of the layer of elongated NPs was 0.9 μm with a FWHM of 0.5 μm.

In order to characterize the NPs orientation in the samples, prior to the SHG experiments, we collected optical absorption spectra using linearly polarized light at two mutually orthogonal polarizations, one parallel (labeled as p) and the other perpendicular (labeled as s) to the plane of incidence. This experiment was performed at three different angles of incidence (0° and ±45°) and an UV-visible spectrophotometer was used to perform the measurements. Figure 11 shows the schematic of these experiments (a and b) and the respective absorption spectra (d and e). Note that Fig. 11 a and b resembles the resulting samples of the two ion implantation process. Note that in Fig. 11 b the particles long axes are tilted 45° with respect to the substrate normal and lay on the xz planes of the laboratory xyz coordinate system. When viewed from the front, the projection of the long axes of the deformed NPs point in the direction we label as x.

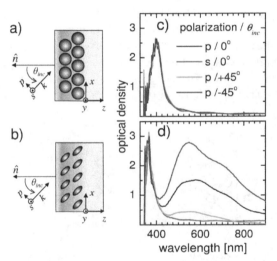

Fig. 11. Schematics for absorption experiments on spherical (a) and elongated (b) Ag-NPs. In the figures, xyz is the laboratory coordinate system; p and s are respectively the parallel and perpendicular linear polarization of the incident beam with respect the plane of incidence; θ_{inc} is the angle of incidence (positive for counterclockwise direction) made by the propagation direction, k, and the surface normal, $\hat{n}$. (c) and (d) are the absorption spectra of spherical and elongated Ag NPs respectively, taken at different input polarization and angle of incidence (as labeled in (c)).

The absorption spectra taken under different angles of incidence have no significant changes for the case of spherical NPs samples. This is shown in Fig. 11c, where a single SP resonance is found at approximately 400 nm. In contrast, from Fig. 11d we can see that for the elongated NPs the absorption spectra depends on both the light polarization and the angle of incidence. At normal incidence the SP resonance of the particles is shifted to lower wavelengths for s-polarization (red line), while the absorption spectra splits into two spectrally separated SP resonances for p-polarization (black line). The shifted resonance at 365 nm obtained with s-polarized light is associated with the short-axis SP and can be explained by the decrease of the NPs size (McMahon et al., 2007) during the second ion implantation process. The resonance at 570 nm obtained with p-polarized light is associated with the long-axis SP and its broadness can be explained by the different NPs sizes formed in the matrix. Note that the 365 nm resonance is also present and presumably invariant in the spectra taken at the three angles of incidence with p-polarized beams, i. e. at −45º (blue line), 0º (black line) and 45º (green line). While in the case of p-polarized light at 45º this result is obvious, since it resembles the case of s-polarized light at 0º, the presence of this band in the other two cases, 0º and −45º, can be attributed to a residual misalignment with respect to the direction of elongation of the particles, or to an actual fraction of smaller spherical NPs remaining in the matrix after the second ion-implantation process. Finally, a strong dependence of the 570 nm SP is obtained with p-polarized light for different angles of incidence being higher when the light propagation is orthogonal to the NPs axes (blue curve). This last result was the criteria used to characterize the NPs orientation used in the SHG experiments.

SHG experiments in the reflection mode with a fixed angle of incidence were conducted using a Ti:Sapphire oscillator as the fundamental beam. The schematic representation of these experiments is shown in Fig. 12a. The laser delivered linearly polarized femto-second pulses with a wavelength centered at 825 nm (pulse width, 88 fs; repetition rate, 94 MHz). The angle of polarization, α, of the fundamental beam was rotated using a $\lambda/2$ wave-plate in order to trace the polar SH dependence of our samples. Using a 50 mm focal length lens, the

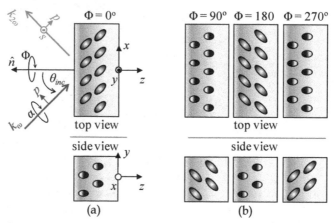

Fig. 12. a) SHG experiment in the reflection mode. In the figure, $\omega, 2\omega$: fundamental and the second harmonic frequencies; p/s:parallel/perpendicular linear polarization of the incident beam with respect the plane of incidence; θ_{inc} : angle of incidence made by the propagation direction, $k\omega$, and the surface normal, $\hat{n}$; α : angle of polarization of the fundamental beam; ϕ: sample rotation angle made by the projection of the NP long axis, ζ, over the xy plane and the fixed plane of incidence contained in xz. (b) Different sample orientations, i. e., elongated NPs orientations with respect the laboratory system, used on SHG experiments.

beam was focused onto the nanocomposite at an angle of incidence $\theta_{inc} = 45°$ with respect the sample surface normal. The reflected SH signal was collected at 90° with respect to the incoming light using a second lens of 30 mm focal length. A color filter and a grating monochromator (not shown) were used to spectrally separate the SH signal from the fundamental light. Finally, the signal was detected via a photomultiplier tube connected to a current/voltage pre-amplifier circuit and a digital oscilloscope. The p-polarized SH (p-SH) and s-polarized SH (s-SH) intensities were also measured using a polarizer cube before signal collection. The sample was mounted on a rotation stage in order to vary the NPs long axis orientation by rotating the angle ϕ. The four different angles used in our experiments are represented in Fig. 12b.

5.3 Results and discussion

Figure 13 shows both the measured and simulated polar dependence of the total (black), p-polarized (red) and s-polarized (blue), SHG intensities obtained for the different 90°-shifted ϕ configurations described in Sec. 5.2. Note that in the experiment, the four configurations produced detectable p- and/or s-SHG signal. We stress out that, in principle, even when the centrosymmetry of the ellipsoidal and spherical NPs is locally disrupted by its surface, the homogeneous polarizing field induces SHG of mutually cancelling polarizations at opposite sides of the circular surface, neglecting then an overall dipolar SHG contribution (Figliozzi et al., 2005; Mendoza & Mochan, 2006). However, we have to bear in mind that no perfect ellipsoids (or spheres) are present in our samples and that the NP size is almost two orders of magnitude smaller that the excitation beam to consider the SH signal as a quadripolar contribution from the NPs bulk. In addition, according to rigorous calculations made by Valencia et. al. (Valencia et al., 2004, 2009), SHG radiation from centrosymmetric infinite cylinders is not symmetric in the back and front surfaces. They find a multi-lobe SHG pattern originated at the cylinder surface, where the angle made from the first SHG scattered lobes in the first surface is more pronounced as the cylinder width is decreased. Seemingly, Bachelier et. al. (Bachelier et al., 2010) modeled both the near-field of the harmonic amplitude and the far-field SH intensity distribution in spherical gold NPs, the two cases show anisotropic radiation patterns arising from the NP surface. Therefore, we attribute this signal to a nonlinear dipolar contribution arising from the NPs surface.

Figure 13a shows, for example, the case when the incident beam is polarized perpendicular to the NP long axis (see Fig. 12a; $\phi=0°$). Here the cross section of the elongated NPs could be considered circularly shaped (from the incident fundamental beam point of view) and therefore no SHG signal is expected according to Eqs. 8-9. In this case, in addition to our argument that no perfect circularly shaped NPs cross section are present in the sample, we attribute this signal to a systematic misalignment in the experiment while rotating the samples as will be shown latter. In contrast, lobes of maximum SH intensity are found at α-angles near the NPs long axes. This is seen in Figs. 13b and d ($\phi=90°$ and $\phi=270°$), respectively, where the s-SH intensity is the main contribution of the total SH signal. Seemingly, the main contribution in Fig. 13c ($\phi=180°$) is the p-SH, instead. Note that Figs. 12b and d are basically mirror images of each other (with y, or $\alpha=90°$, as the symmetry axis) with total SHG maxima at ~ 115° and ~ 290°, for b, and ~ 65° and ~ 255° for d, respectively; the counterpart p-SH intensity has practically no contribution. Otherwise, from Fig. 13c we can see that both the p- and s-SH intensities contribute to the total SH when the fundamental beam is perpendicular to the NPs long axes (Fig. 12b; $\phi=180°$).

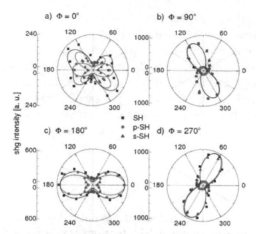

Fig. 13. Experimental and simulated SHG polar dependence of elongated Ag NPs, for the four different sample orientations (see Fig. 4) ϕ =0° (a), ϕ =90° (b), ϕ =180° (c) and ϕ =270° (d), respectively. In all plots the experimental total, p-polarized and s-polarized, SH are denoted by black squares, red circles and blue triangles, respectively. While the simulated SHG intensities are denoted using the same color convention in solid lines. An offset has been intentionally added in all plots to see the lower signal (see p -SH in b and d, for example).

The simulated polar traces are in a good agreement with the experimental results; this can be seen also in Fig. 13. Table 1 shows specific values of the parameters r and a, used in Eqs. 8-9, that best fitted with the experimental data, where, ϕ_{sim}, stands for the simulated value of ϕ. The simulated data revealed two extra peaks in between the s-SH maxima ($\sim$ 25 times lower), in Figs. 13b and d, respectively, and their values are found also in Tab. 1. Seemingly, two extra peaks are also found in Fig. 13c, but at much smaller values ($\sim$ 300 times less). Note also that the values of r were fitted in the range of 1.4<1/r<3.5 for cases 2–4 (see table), confirming that there is a stronger hyperpolarizability response for fields oscillating along the NPs long axis (i.e. $\beta_{\zeta\zeta\zeta} > \beta_{\xi\xi\zeta}$). These values are very similar to the values obtained in synthetic films consisting of helical (PBLG) macromolecules (Knoesen et al., 2004) and native fibrillar collagen (Stoller et al., 2003), their hyperpolarizability are reported to be within an order of magnitude of that of crystalline quartz (Freund et al., 1986).

Case	ϕ	1/r	a	ϕ_{sim}	$\left\|\chi^{(2)}_{eff,pa}\right\|^2_{max}$	$\left\|\chi^{(2)}_{eff,sa}\right\|^2_{max}$
1.	0°	1/1.41	16	176°	0°, 180°	45°, 135°
2.	90°	3.5	52	87°	25°, 115° 205°, 295°	65°, 155° 245°, 335°
3.	180°	1.4	34	180°	0°, 90° 180°, 270°	45°, 135° 225°, 315°
4.	270°	3.5	57	271°	66°, 156° 246°, 336°	26°, 116° 206°, 295°

Table 1. Values of parameters, r and a, and resulting SH maxima positions, to simulate the SHG experiments on elongated Ag NPs.

Note that simulating the different cases shown in Figs. 13b-c with the same parameters (r and a) would result in obtaining higher SHG maxima in Fig. 13c than in Figs. 13b and d. However, in the experiment we obtained less SHG signal in Fig. 13d and we attribute this result to the presence of less NPs within the point spread function of the fundamental beam and/or a minor hyperpolarizability value. The inhomogeneous NPs distribution in the composite film makes extremely challenging maintaining the same irradiated area in the experiment while rotating the sample. As a consequence, the value of the parameters a and r used to fit the experimental data were different. As can be seen from Tab. 1, in this experiment the parameters used in Fig. 13c resulted to be smaller than the respective parameter values used to fit the experimental data of Figs. 13b and d. Otherwise, it is interesting to note that the experimental case at $\phi = 0^o$ (Fig. 13a) is reproduced for angles ϕ_{sim} close to 180° and $r= 1.41$, this simply indicates that we can simulate this result by assuming stronger hyperpoplarizability responses for fields oscillating perpendicular to the NPs long axis (*i.e.* $\beta_{\zeta\zeta\zeta} < \beta_{\xi\xi\zeta}$). The asymmetric polar dependence was obtained using ϕ_{sim} confirming that our experimental results are most probable due to misalignment. Note also that both the p- and s-SHG intensities are comparable in magnitude, while for the case shown in Fig. 13c the p-SH intensity is ~5 times larger than s-SH.

In order to be sure that the results correlate indeed with the known (simulated) structure of the elongated NPs, SHG experiments were also made in spherical NPs. Figure 14 shows SH signal from samples with embedded spherical NPs. The total SHG (black) presents nearly isotropic polar trace where p-SH (red) is the maximum signal contribution. The s-SH intensity (blue) also contributes but the signal is ~ 10 times lower than the p-SH counterpart. It presents a characteristic shape with maxima at 45°, 135°, 225° and 315°. We found the same dependence for different ϕ values 0° and 90° (Fig. 14a and b), respectively, indicating that the obtained polar traces are a characteristic of the spherical NPs. The fundamental and SH beams spectra for elongated (black) and spherical (red) NPs, with the respective absorption spectra (dotted curves) are also shown in Fig. 15a. The absorption traces indicate that for elongated NPs the SHG may be enhanced since the fundamental is close to resonance with the SP broad band (black dotted line) at 570 nm (see also blue curve on Fig. 11d). Note, however, that the SHG suffers also absorption of about the same optical density reducing the signal. In contrast, the SP resonance of spherical NPs (red dotted line in Fig. 15a; which is the red solid line in Fig. 11c) is far away of the fundamental wavelength and therefore no enhancement effect is expected. In addition since the NPs sizes are small compared to the fundamental wavelength, this suggest that the SHG in spherical NPs samples arises mainly from the electric dipole surface contribution, owing to the actual non perfect spherical shape of the particles (Nappa et al., 2005), and must be large enough to be detectable even after being absorbed with an optical density of ~3. Otherwise, the quadratic dependence with respect to the input power obtained in both types of samples, Fig. 15b (for elongated NPs) and c (for spherical ones), indicates the typical coherent nature of nonlinear scatterers. In particular, the result obtained in spherical NPs indicates that the SH observed is not due to grating effects such as the hyper Rayleigh scattering (HRS), where incoherent SHG is produced for which a typical linear dependence with respect to the input power is observed.

Our results are in accordance with earlier SHG experiments performed by Podlipensky *et. al.* (Podlipensky et al., 2003) on elongated Ag NPs. The main differences with respect to our

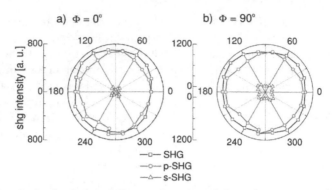

Fig. 14. SHG signal of spherical Ag NPs as a function of the polarization angle, α, obtained for two different sample orientations (see Figure 2): $\phi=0^o$ (a) and $\phi=90^o$ (b). In the plots, the total (opened squares, black), p-polarized (opened circles, red) and s-polarized (opened triangles, blue) SHG intensities are shown.

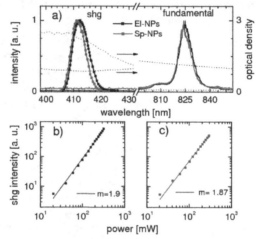

Fig. 15. (a) Fundamental laser spectra used in the SHG experiments and SHG spectra obtained for spherical and elongated NPs. In all plots, curves in black stands for elongated NPs while curves in red for spherical NPs. SHG signal as a function of the fundamental input power for elongated (b) and spherical (c) Ag-NPs, respectively. Here, solid squares denote the experimental data while continuous lines indicate the fitted curves. m: is the slope obtained from the linear fitting.

experiments are that this group obtained equivalent intensities for both p- and s-SH signal and no measurable signal for spherical NPs was detected. Their SH experiments were performed with the fundamental beam at an angle of incidence close to the surface normal, 15° (SH collected in transmission), and the NPs long axes aligned along the surface normal. We consider that such experimental arrangement is close to the case shown in Fig. 12a, since the direction of the fundamental beam is also close to the direction of the NPs long axis and comparable intensities are obtained for p and s-SH. As discussed above we believe that we are able to detect SH from spherical NPs due to the fact that we have smaller NPs sizes (at least 10 times smaller) with respect to Podlipensky's samples. Note that Eqs. 8-9 do not explain this dependence, since they were obtained considering a hyperpolarizability tensor with cylindrical symmetry, however, it is interesting to note that our experimental results can be explained using analytical expressions obtained by Dadap *et. al.* (Dadap et al., 1999) to describe SH Rayleigh Scattering from spheres of centrosymmetric material, where the intensities for vertical and horizontal SH are given by $I_{pa} \propto |a_1|^2$ and $I_{sa} \propto |a_2|^2 \sin 2\alpha$, respectively. In these expressions, p and s, stands for the horizontal and vertical polarization of the harmonic generated signals, respectively, α, is the fundamental input polarization, and, a_1 and a_2, are complex numbers related to the pure effective dipole contribution and quadruple contribution, respectively. For p-SHG the intensity is constant, independent of the input polarization angle α, while for s-SHG intensity the signal is maximum at $\alpha = (2n-1)45°$ and vanishes at $\alpha = (n-1)90°$, with n =integer. Fig. 6b has been intentionally altered in order to see such s-SHG polar behavior. In addition, being the p-SH intensity higher with respect to the s-SH counterpart in our experiments (then $a_1 >> a_2$), strongly supports our assumption that the SHG is dominated by dipolar contributions arising from the surface of each non-perfectly spherical Ag-NP and having a quadratic response with respect to the input power (observed in Fig. 15c) confirms their coherent summation (Rocha-Mendoza et al., 2011).

6. Conclusions

The third-order nonlinear results here presented allows a better understanding of the nonlinear optical response of plasmonic nanocomposites, based on the electronic transitions occurred in the embedded NPs. For the studied wavelengths, the response depends on the incident irradiance, mainly because of the hot-electrons contribution. Regarding anisotropic plasmonic nanocomposites, for the wavelengths used in this work, the analysis may be done considering only one plasmon resonance at a time. This fact allows making the same considerations as for the isotropic case, and, for applications, by varying wavelength, polarization and angular position of the sample, one can switch the sign of the nonlinear optical response of these nanocomposites, which gives a wide picture of possible applications of these nanocomposites into the plasmonics realm.

Second-harmonic generation from composites containing randomly distributed but aligned elongated silver nanoparticles has been presented and modeled as a coherent summation of the microscopic hyperpolarizability associated to each NP. Our experimental data suggest that the origin of the hyperpolarizability, in both elongated and spherical NPs, can be attributed mainly to a surface nonlinear contribution of each non-perfect ellipsoidal or spherical Ag-NP.

7. Acknowledgments

The authors wish to acknowledge the technical assistance of K. López, F. J. Jaimes, J. G. Morales and E. J. Robles-Raygoza. We would like to thank Dr. Eugenio Méndez for useful discussions. Finally, we also acknowledge the financial support from PAPIIT-UNAM through grants IN108510 and IN103609; from CONACyT through grants 80019 and 102937, from ICyT-DF through grant PICCT08-80, and (I. Rocha-Mendoza) from UC-MEXUS/CONACyT under collaborative research programs.

8. References

Aktsipetrov, O. A.; Elyutin, P. V.; Nikulin, A. A. & Ostrovskaya, E. A. (1995). Size effects in optical second-harmonic generation by metallic nanocrystals and semiconductor quantum dots: The role of quantum chaotic dynamics. *Phys. Rev. B*, 51, 17591–17599.

Bachelier, G.; Butet, J.; Russier-Antoine, I.; Jonin, C.; Benichou, E. & Brevet, P.-F. (2010). Origin of optical secondharmonic generation in spherical gold nanoparticles: Local surface and nonlocal bulk contributions. *Phys. Rev. B*, 82, 235403.

Barnes, W. L.; Dereux, A. & Ebbesen, T. W. (2003). Surface plasmon subwavelength optics. *Nature*, 424, 824–830.

Boyd, R. (2008). *Nonlinear optics* (3rd. Edition), Academic Press, USA.

Brevet, P-F. (2011) Second Harmonic Generation, *Comprehensive Nanoscience and Technology*, G. A. Wurtz, R.J. Pollard and A.V. Zayats, eds., 351–381, Elsevier.

Campagnola, P. J.; & Loew, L. M. (2003). Second-harmonic imaging microscopy for visualizing biomolecular arrays in cells, tissues and organisms. *Nature Biotech.*, 21, 1356–1360.

Dadap, J. I.; Shan, J.; Eisenthal, K. B. & Heinz, T. F. (1999). Second-Harmonic Rayleigh Scattering from a Sphere of Centrosymmetric Material. *Phys. Rev. Lett.*, 83, 4045–4048.

Dadap, J. I.; Shan, J. & Heinz, T. F. (2004). Theory of optical second-harmonic generation from a sphere of centrosymmetric material: small-particle limit. *J. Opt. Soc. Am. B*, 21, 1328-1347.

Fernández-Hernández, R. C.; Gleason-Villagrán, R.; Torres-Torres, C.; Cheang-Wong, J. C.; Crespo-Sosa, A.; Rodríguez-Fernández, L.; López-Suárez, A.; Rangel-Rojo, R.; L; Oliver, A. & Reyes-Esqueda, J. A. (2011). Nonlinear optical spectroscopy of isotropic and anisotropic metallic nanocomposites. *J. Phys.: Conference Series*, 274, 012074.

Figliozzi, P.; Sun, L.; Jiang, Y.; Matlis, N.; Mattern, B.; Downer, M. C.; Withrow, S. P.; White, C.W.; Mochán, W. L. & Mendoza, B. S. (2005). Single-Beam and Enhanced Two-beam second-harmonic generation from silicon nanocrystals by use of spatially inhomogeneous femtosecond pulses. *Phys. Rev. Lett.*, 94, 047401.

Freund, I.; Deutsch, M. & Sprecher, A. (1986). Connective tissue polarity. Optical second-harmonic microscopy, crossed-beam summation, and small-angle scattering in rat-tail tendon. *Biophys. J.*, 50, 693–712.

Gallet, S.; Verbiest, T. & Persoons, A. (2003). Second-order nonlinear optical properties of nanocrystalline maghemite particles. *Chem. Phys. Lett.*, 378, 101–104.

Hache, F.; Ricard, D. & Flytzanis, C. (1986). Optical nonlinearities of small metal particles: surface-mediated resonance and quantum size effects. *J. Opt. Soc. Am. B*, 3, 1647-1655.

Hache, F.; Ricard, D.; Flytzanis, C. & Kreibig, U. (1988). The optical kerr effect in small metal particles and metal colloids: the case of gold. *Appl. Phys. A: Mater. Sci. Process.*, 47, 347-357.

Hirose, C.; Akamatsu, N. & Domen, K. (1992). Formulas for the Analysis of the Surface SFG Spectrum and Transformation Coefficients of Cartesian SFG Tensor Components. *Appl. Spectrosc.*, 46, 1051–1072.

Inouye, H; Tanaka, K.; Tanahashi, I.; Hattori, T. & Nakatsuka, H. (2000). Ultrafast Optical Switching in a Silver Nanoparticle System. *JJAP*, 39, 5132–5133.

Karthikeyan, B.; Anija, M.; Suchand Sandeep, C. S.; Muhammad Nadeer, T. M. & Philip, R. (2008). Optical and nonlinear optical properties of copper nanocomposite glasses annealed near the glass softening temperature. *Opt. Commun.*, 281, 2933-2937.

Khosrofian, J. M. & Garetz, B. A. (1983). Measurement of a Gaussian laser beam diameter through the direct inversion of knife-edge data. *Appl. Opt.*, 22, 3406-3410.

Knoesen, A.; Sakalnis, S.; Wang, M.; Wise, W. D.; Lee, N. & Frank, C. W. (2004). Sum-frequency spectroscopy and imaging of aligned helical polypeptides. *IEEE J. Sel. Top. Quantum Electron.*, 10, 1154–1163.

Leray, A. ; Leroy, L. ; Le Grand, Y.; Odin, C. ; Renault, A.; Vi, V.; Roude, D.; Mallegol, T.; Mongin, O. ; Werts, M. H. V. & Blanchard-Desce, M. (2004). Organization and Orientation of Amphiphilic Push-Pull Chromophores Deposited in Langmuir-Blodgett Monolayers Studied by Second Harmonic Generation and Atomic Force Microscopy. *Langmuir*, 20, 8165–8171.

Liu, X.; Guo, S.; Wang, H. & Hou, L. (2001). Theoretical study on the closed aperture Z-scan curves in the materials with nonlinear refraction and strong nonlinear absorption. *Opt. Commun.*, 197, 431-437.

Matsui, I. (2005) Nanoparticles for Electronic Device Applications: A Brief Review, *JCEJ*, 38, 535–546.

McMahon, M. D.; Ferrara, D.; Bowie, C. T.; Lopez, R. & Haglund Jr., R. F. (2007). Second harmonic generation from resonantly excited arrays of gold nanoparticles. *Appl. Phys. B*, 87, 259–265.

Mendoza, B. S. & Mochán, W. L. (2006). Second harmonic surface response of a composite," *Opt. Mat.*, 29, 1–5.

Nappa, J.; Revillod, G.; Russier-Antoine, I.; Benichou, E.; Jonin, C. & Brevet, P-F. (2005). Electric dipole origin of the second harmonic generation of small metallic particles. *Phys. Rev. B*, 71, 165407.

Noguez, C. (2007). Surface plasmons on metal nanoparticles: the influence of shape and physical environment. *J. Phys. Chem. C*, 111, 3806-3819.

Oliver, A.; Reyes-Esqueda, J. A.; Cheang-Wong, J. C.; Román-Velázquez, C. E.; Crespo-Sosa, A.; Rodríguez-Fernández, L.; Seman, J. A. & Noguez, C. (2006). Controlled anisotropic deformation of Ag nanoparticles by Si ion irradiation. *Phys. Rev. B*, 74, 245425.

Podlipensky, A.; Lange, J.; Seifert, G.; Graener, H. & Cravetchi I. (2003). Second-harmonic generation from ellipsoidal silver nanoparticles embedded in silica glass. *Opt. Lett.* 28, 716–718.

Psilodimitrakopoulos, S.; Santos, S. I. C. O.; Amat-Roldan, I.; Thayil, A. K. N.; Artigas, D. & Loza-Alvarez, P. (2009). *In vivo*, pixel-resolution mapping of thick filaments' orientation in nonfibrilar muscle using polarization-sensitive second harmonic generation microscopy," *J. Biomed. Opt.*, 14, 014001.

Rangel-Rojo, R.; McCarthy, J.; Bookey, H. T.; Kar, A. K.; Rodríguez-Fernández, L.; Cheang-Wong, J. C.; Crespo-Sosa, A.; Lopez-Suarez, A.; Oliver, A.; Rodríguez-Iglesias, V. & Silva-Pereyra, H. G. (2009). Anisotropy in the nonlinear absorption of elongated silver nanoparticles in silica, probed by femtosecond pulses. *Opt. Commun.*, 282, 1909–1912.

Rangel-Rojo, R; Reyes-Esqueda, J. A.; Torres-Torres, C.; Oliver, A.; Rodríguez-Fernández, L.; Crespo-Sosa, A.; Cheang-Wong, J. C.; McCarthy, J.; Bookey, H. T. & Kar, A. K. (2010). Linear and nonlinear optical properties of aligned elongated silver nanoparticles embedded in silica, In: *Silver Nanoparticles*, David Pozo Perez eds., 35–62, InTech, http://www.intechopen.com/articles/show/title/linear-and-nonlinear-optical-propertiesof-aligned-elongated-silver-nanoparticles-embedded-in-silica.

Reyes-Esqueda, J. A.; Torres-Torres, C.; Cheang-Wong, J. C.; Crespo-Sosa, A.; Rodríguez-Fernández, L.; Noguez, C. & Oliver, A. (2008). Large optical birefringence by anisotropic silver nanocomposites. *Optics Express*, 16, 710-717.

Reyes-Esqueda, J. A.; Rodríguez-Iglesias, V.; Silva-Pereyra, H. G.; Torres-Torres, C.; Santiago-Ramírez, A.-L.; Cheang-Wong, J. C.; Crespo-Sosa, A.; Rodríguez-

Fernández, L.; López-Suárez, A. & Oliver, A. (2009). Anisotropic linear and nonlinear optical properties from anisotropy-controlled metallic nanocomposites. *Optics Express*, 17, 12849-12868.

Rocha-Mendoza, I.; Yankelevich, D. R.; Wang, M.; Reiser, K. M.; Frank, C. W. & Knoesen, A. (2007). Sum Frequency Vibrational Spectroscopy: The Molecular Origins of the Optical Second-Order Nonlinearity of Collagen. *Biophys. J.*, 93, 4433-4444.

Rocha-Mendoza, I; Rangel-Rojo, R.; Oliver, A. & Rodríguez-Fernández, L. (2011). Second-order nonlinear response of composites containing aligned elongated silver nanoparticles. *Optics Express* 19, 21575-21587.

Rodríguez-Iglesias, V.; Silva-Pereyra, H. G.; Cheang-Wong, J. C.; Reyes-Esqueda, J. A.; Rodríguez-Fernández, L.; Crespo-Sosa, A.; Kellerman, G. & Oliver, A. (2008). MeV Si ion irradiation effects on the optical absorption properties of metallic nanoparticles embedded in silica. *Nuc. Instrum. Meth. B*, 266, 3138-3142.

Rodríguez-Iglesias, V.; Silva-Pereyra, H. G.; Torres-Torres, C.; Reyes-Esqueda, J. A.; Cheang-Wong, J. C.; Crespo-Sosa, A.; Rodríguez-Fernández, L.; López-Suárez, A. & Oliver, A. (2009). Large and anisotropic third-order nonlinear optical response from anisotropy-controlled metallic nanocomposites. *Opt. Commun.*, 282, 4157-4161.

Ryasnyansky, A.; Palpant, B.; Debrus, S.; Khaibullin, R. I. & Stepanov, A. L. (2006). Nonlinear optical properties of copper nanoparticles synthesized in indium tin oxide matrix by ion implantation. *J. Opt. Soc. Am. B*, 23, 1348-1353.

Sheik-Bahae, M.; Said, A. A.; Wei, T. H. & van Stryland, E. W. (1990). Sensitive Measurement of Optical Nonlinearities Using a Single Beam. *IEEE J. Quant. Electron.*, 26, 760-769.

Shen, Y. R. (1989). Surface properties probed by second-harmonic and sum-frequency generation. *Nature*, 337, 519-525.

Silva-Pereyra, H. G.; Arenas-Alatorre, J.; Rodríguez-Fernández, L.; Crespo-Sosa, A.; Cheang-Wong, J. C.; Reyes-Esqueda, J. A. & Oliver, A. (2010). High stability of the crystalline configuration of Au nanoparticles embedded in silica under ion and electron irradiation. *J. Nanopart. Res.*, 12, 1787-1795.

Stoller, P.; Celliers, P. M.; Reiser, K. M. & Rubenchik, A. M. (2003). Quantitative second-harmonic generation microscopy in collagen. *Appl. Opt.*, Vol. 42, No. 25, 5209-5219.

Tominaga, J.; Mihalcea, C.; Buchel, D.; Fukuda, H.; Nakano, T.; Atoda, N.; Fuji, H. & Kikukawa, T. (2001). Local plasmon photonic transistor. *Appl. Phys. Lett.*, 78, 2417-2419.

Torres-Torres, C.; Reyes-Esqueda, J. A.; Cheang-Wong, J. C.; Crespo-Sosa, A.; Rodríguez-Fernández, L & Oliver, A. (2008). Optical third-order nonlinearity by nanosecond and picosecond pulses in Cu nanoparticles in ion-implanted silica. *J. Appl. Phys.*, 104, 014306.

Valencia, C. I.; Méndez, E. R. & Mendoza, B. S. (2004). Second-harmonic generation in the scattering of light by an infinite cylinder. *J. Opt. Soc. Am. B*, 21, 36-44.

Valencia, C. I.; Méndez, E. R. & Mendoza, B. S. (2009). Weak localization effects in the second-harmonic light scattered by random systems of particles. *Opt. Commun.*, 282, 1706-1709.

Wang, K.; Long, H.; Fu, M.; Yang, G. & Lu, P. (2010). Size-related third order optical nonlinearities of Au nanoparticle arrays. *Optics Express*, 18, 13874-13879.

Zheludev, N. I. & Emelyanov, V. I. (2004). Phase matched second harmonic generation from nanostructured metallic surfaces. *J. Opt. A: Pure App. Opt.*, 6, 26-28.

Zhuang, X.; Miranda, P. B.; Kim, D. & Shen, Y. R. (1999). Mapping molecular orientation and conformation at interfaces by surface nonlinear optics. *Phys. Rev. B*, 59, 12632-12640.

3

Stimulated Raman Scattering in Quantum Dots and Nanocomposite Silicon Based Materials

M. A. Ferrara*, I. Rendina and L. Sirleto

National Research Council-Institute for Microelectronics and Microsystems, Napoli, Italy

1. Introduction

Stimulated Raman Scattering (SRS) is one of the first discovered nonlinear optical effects: a pump laser beam enters a nonlinear medium and spontaneous generation and amplification lead to a beam at a frequency different from the pump. SRS is dependent on the pump intensity and on a gain coefficient g, which depends on material scattering efficiency: the larger the spontaneous scattering efficiency of materials is, the higher the Raman gain for a given intensity is obtained. As a general rule, there is a trade-off between gain and bandwidth in all laser gain materials: line-width is bought at the expense of peak gain. Of course, this is true for bulk solids, but the question is what happens at nanoscale?

Nonlinear optics at nanoscale is a recent fascinating research field. Stimulated Raman scattering in electrons-confined and photons-confined materials is of great importance from both fundamental and applicative point of view. Concerning the fundamental one, there have been a number of investigations both experimental and theoretical, but the question is still "open", while from an applicative point of view, there are some important prospective, for example to realize micro/nano source, with improved performances.

In this chapter, experimental investigations of stimulated Raman scattering in silicon quantum dots and in silicon nanocomposite are reported. Two Raman amplifiers are realized and amplifications due to stimulated Raman scattering are measured. For both of them, a significant enhancement of Raman gain and a significant reduction in threshold power are demonstrated. Our findings indicate that nanostructured materials show great promise for Si-based Raman lasers.

2. Basis on Stimulated Raman Scattering and Raman amplifier

SRS belongs to a class of nonlinear optical processes that can be called quasi-resonant. Although none of the fields is in resonance with the vibrations in the lattice of the medium (optical phonons), the difference between the pump and generated beam equals the transition frequency. SRS is used in tunable laser development, high energy pulse compression, etc.

SRS can be obtained by irradiating a solid with two simultaneous light sources: a light wave at frequency ω_L (the pump laser wave) and a light wave at frequency $\omega_S = \omega_L - \omega_v$ (the

---

* Corresponding Author

Stokes Raman wave), where $\hbar\omega_v$ corresponds to a vibrational energy (Fig. 1). The pump causes the molecular vibration and thereby impress frequency sidebands (Stokes and anti-Stokes). From the other side, Stokes wave at frequency ω_S can beat the laser to produce a modulation of the total intensity that coherently excites the molecular oscillation at the frequency $\omega_v = \omega_L - \omega_S$. These two process due to pump and Stokes waves, reinforce one another in the sense that the pump effect leads to a stronger Stokes wave, which in turn leads to a stronger molecular vibration (Boyd, 2003).

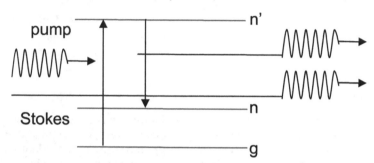

Fig. 1. Stimulated Raman Scattering.

In the steady-state (no pump depletion) regime of SRS, the intensity of the output Stokes radiation is expressed by (Shen & Bloembergen , 1965)

$$I_S(L) = I_S(0) \cdot \exp(I_P(0)gL) \tag{1}$$

where I_{S0} is the intensity of the input Stokes radiation (Stokes seed), I_S is the intensity of the output Stokes radiation, I_P is the intensity of the pump radiation, g is the Raman gain coefficient, and L is the effective length. Assuming no losses at the Stokes frequency, the value of the gain coefficient g can be obtained by fitting Eq. (1), which is readily transformed into

$$SRS = 10 \cdot \log_{10}\left(\frac{I_S(L)}{I_S(0)}\right) = 4.34 \cdot gLI_P(0) \tag{2}$$

where $I_P = P/A$ with P as the power incident onto the sample and A as the effective area of pump beam. Since the sample transparent to the incident light, L is taken to be equal to the thickness of the sample along the path of the incident light.

Raman lasing can be achieved by using the Stimulated Raman Scattering phenomenon, which permits, in principle, the amplification in a wide interval of wavelengths, from the ultraviolet to the infrared.

Fused silica has been, for the past century, the key material used for long and short haul transmission of optical signals, because of its good optical properties and attractive figure of merit (i.e. trade-off between Raman gain and losses). A breakthrough in fiber optics communications was achieved with the reduction of the water absorption peak at 1400 nm, which opened up the available communication range to span from 1270 to 1650 nm, corresponding to about 50 THz bandwidth (Rivero et al., 2004). This dramatic increase in bandwidth rules out the use of existing Er doped fiber amplifiers, leaving Raman gain as the main mechanism for future amplification needs. However, the main disadvantage of the current silica fiber amplifiers is the limited usable bandwidth for Raman amplification (5 THz, approx. 150 cm^{-1}).

On the other hand, in the past few years several strategies have been developed to engineer efficient light sources and amplifiers in silicon-based materials (Pavesi & Lockwood, 2004; Soref, 2006), with the aim to demonstrate a convenient path to monolithic integration of optical and electronic devices within the mainstream Si technology. In particular, Raman amplification is an interesting approach for optical amplification, because it is only restricted by the pump wavelength and Raman active modes of the gain medium (Islam, 2002; Mori et al., 2003). Light amplification by stimulated Raman scattering in silicon waveguides has been recently demonstrated, despite intrinsic limitations related to the nature of the bulk Si materials have been pointed out (Jalali et al., 2006; Dekker et al., 2007).

Raman scattering in bulk silicon was studied as early as 1965 (Russell, 1965). Using a helium–neon laser with an output wavelength of 0.6328 μm, backward Raman scattering from silicon was measured, and it was found that the Raman scattering efficiency in silicon was 35 times larger than that for diamond (Russell, 1965). In 1970 (Ralston & Chang, 1970), more detailed Raman experiment using a YAG:Nd laser having a wavelength of 1.064 μm was performed. Both spontaneous and stimulated Raman scattering efficiency was characterized experimentally (Ralston & Chang, 1970). The observed Raman frequency downward shift of 15.6 THz corresponds to optical phonon energy of silicon at the center of the Brillouin zone (Hart et al., 1970; Temple & Hathaway, 1973). The first-order resonance, which is of primary importance here, has a full-width at half-maximum of 105 GHz (Temple & Hathaway, 1973). This imposes a maximum information bandwidth of approximately 105 GHz that can be amplified. The Raman linewidth becomes broader when a broadband pump is used. These experiments has the merit to prove that silicon has a relatively strong Raman scattering efficiency (four orders of magnitude higher than that for silica) (Agrawal, 1995; Claps et al., 2002; Claps et al., 2003).

2.1 SRS in silicon waveguide

SRS has been exploited in optical fibers to create amplifiers and lasers. However, several kilometers of fiber is typically required to create a useful device, suggesting that the approach is not applicable in integrated devices. Often overlooked was the fact that the gain coefficient for SRS in silicon is approximately 10^3–10^4 times higher than that in silica fiber. Additionally, owing to the large refractive index, silicon waveguides can confine the optical field to an area that is approximately 100-1000 times smaller than the modal area in a standard single-mode optical fiber, resulting in proportionally higher Raman gain (Claps et al., 2002). When combined, these facts make it possible to observe SRS over the interaction lengths encountered on a chip (Claps et al., 2002; Claps et al., 2003; OSA Press Room Editorial).

As silicon is transparent in 1.3-1.6 μm (optical communication band), Raman scattering in silicon waveguides in such a wavelength range has attracted a great deal of interest in the past couple of years (Claps et al., 2002; Rong at al., 2005).

The use of SRS in silicon waveguides was proposed in 2002 as a means to realize silicon amplifiers and lasers (Claps et al., 2002; OSA Press Room Editorial). This was followed by demonstration of stimulated emission (Claps et al., No. 15, 2003) and Raman wavelength conversion (Claps et al., No. 22, 2003) in silicon waveguides in 2003. Shortly afterward, the approach led to the demonstration of the first silicon laser in 2004: a device that operated in the pulsed mode (Boyraz & Jalali, 2004; Nature News Editorial) followed by demonstration of continuous-wave (CW) lasing in 2005 (Rong et al., No. 7027, 2005).

Pulsed lasing has been achieved using a ring cavity formed by an 8-m-long optical fiber and silicon as the gain medium via SRS (Boyraz & Jalali, 2004). The slope efficiency, which is described by the ratio of the output peak power and the input peak pump power, obtained was 8.5%.

An all-silicon Raman CW laser has been demonstrated using centimeter-size silicon waveguide (Rong et al., No. 7023, 2005; Rong et al., No. 7027, 2005). Recently, the waveguide was replaced by a ring cavity witch have a total length of 3 cm and a bend radius of 400 μm (Rong et al., 2006). A slope efficiency of ~10% was obtained.

2.2 Problems in silicon waveguide Raman amplifiers

In two-photon absorption (TPA) process an electron absorbs two photons from the laser (Fig. 2) at approximately the same time (or within less than a nanosecond) and achieves an excited state that corresponds to the sum of the energy of the incident photons. This is a nonlinear process, occurring with significant rates only at high optical intensities, because the two-photon absorption coefficient is proportional to the optical intensity of pump laser. TPA can be eliminated entirely choosing a material for which the lowest-lying excited state lies more than $2\hbar\omega$ above the ground state (Boyd, 2003).

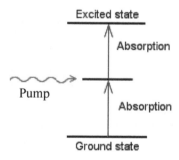

Fig. 2. Two Photon Absorption.

As a multiphoton process, TPA is a competitive effect of stimulated Raman scattering. In particular the two-photon absorption (TPA) reduces the efficiency of SRS phenomena.

The main challenge in silicon Raman laser is the loss caused by the free carriers that are generated via TPA (Liang & Tsang, 2004; Claps et al., 2004). By determining the steady-state density of generated carriers, and hence the magnitude of the pump-induced loss, the recombination lifetime is the central parameter in Raman as well as other semiconductor nonlinear optical devices. Due to the presence of interface states at the boundary between the top silicon and the buried oxide layer, the recombination lifetime in SOI is much shorter than that in a bulk silicon sample with comparable doping concentration. This effect depends on the method used for preparation of the SOI wafer and the film thickness, with measured and expected values ranging between 10 and 200 ns (Mendicino, 1998; Freeouf & Liu, 1995). In SOI waveguides, the lifetime is further reduced to a few nanoseconds, or even below in the case of submicrometer waveguides, due to the recombination at the etched waveguide facets and, in the case of rib waveguides, due to diffusion into the slab regions (Dimitropoulos et al., Vol. 86, 2005; Espinola et al., 2004). By introducing midgap states through high-energy irradiation and gold or platinum doping, the lifetime can be further reduced. The carrier density can also be reduced by using a reverse-bias p-n junction to sweep the carriers out (Liang & Tsang, 2004; Claps et al., 2004), and CW gain using this approach has been reported (Liu et al., 2004; Fathpour et al., 2006). However, free-carrier screening of the junction electric field, a phenomenon reminiscent of high power saturation in photodetectors, limits the usefulness of this technique to modest pump intensities (Dimitropoulos et al., Vol. 87, 2005). Furthermore, the diode results in electrical power being dissipated on the chip.

Being optically pumped, it is unlikely that the Raman laser will play a role in optical interconnects; on the other hand, by compensating for coupling and propagation losses, a Raman amplifier can have an impact. However, the device length will have to be drastically lower than the centimeter-length device demonstrated so far. At an intensity of 100 MW/cm^2 (100 mW pump coupled into a 0.1μm^2 waveguide) and with silicon's Raman gain coefficient of ~20 cm/GW, the gain will be ~1 dB/mm. Such a modest gain per unit length creates a challenge for miniaturization of Raman amplifiers, a prerequisite for their integration with silicon VLSI-type circuits. The long waveguide length will not be an issue if the device is used as a stand-alone discrete amplifier (similar to the role played by the EDWA).

2.3 Silicon nanostructures

Taking into account intrinsic limitations related to the nature of the bulk Si materials, that is the narrow-band (105 GHz) of stimulated Raman gain, the small Raman effect and the competing nonlinear effect of TPA, which reduces the efficiency of SRS, the investigation of new materials possessing both large Raman gain coefficients and broader spectral bandwidth than fused silica and/or silicon is becoming mandatory in order to satisfy the increasing telecommunications demands.

Nanostructured silicon has generated large interest in the past decades as a promising key material to establish a Si-based photonics. An accurate knowledge of both linear and nonlinear optical properties of these structures is crucial for the conception and design of highly efficient photonic structures and for the control of their performance (Pavesi & Lockwood, 2004).

Strong enhancement (~10^3) of the spontaneous Raman scattering was reported from individual silicon nanowires and nanocones as compared with bulk Si (Cao et al., 2006). The observed enhancement was diameter, excitation wavelength, and incident polarization state dependent. The observed increase in Raman-scattering intensity with decreasing diameter in this system was explained in terms of structural resonances in the local field similar to Mie scattering from dielectric spheres.

We note that SRS from spherical droplets and microspheres, with diameters 5÷20 μm, has been observed using both pulsed and continuous wave probe beams (Spillane et al., 2002).

Except for report of SRS from individual single walled carbon nanotubes (Zhang et al., 2006), and the observation of SRS from semiconductor nanowires (Jian Wu et al., 2009), we find no other evidence for this important nonlinear optic effect in nanostructured materials.

3. Broadening and tuning of Spontaneous Raman Scattering in silicon nanocrystals

In this paragraph, some advantages of Raman approach in silicon nanocrystals with respect to silicon are theoretically and experimentally demonstrated. In order to provide theoretical basis for these results, phonon confinement model is briefly introduced. After that, according to this model, we discuss two significant improvements of this approach: the broadening of Raman scattering and the tuning of Stokes shift in Si-nc with respect to silicon.

When the size of the particle reduce to the order of nm, the wave function of optical phonons will non longer be a plane wave. The localization of wave function leads a relaxation in the selection rule of wave vector conservation. Not only the phonons with zero wave vector q=0, but also those with q>0 take part in the Raman scattering process, resulting in the red shift of the peak position and the broadening of the peak width. A quantitative model, developed by Campbell and Fauchet, calculates that the peak position mainly depends on the number of atoms included in a cluster, while the width of spectra depends on the shape of crystallites (Ritcher et al, 1981; Campbell & Fauchet, 1986).

Silicon nanocrystals sample can be modelled as an assembly of quantum dots and the phonon confinement is three dimensional. The weight factor of the phonon wave function is chosen to be a Gaussian function as follows:

$$W(r,L) = \exp\left(-\frac{8\pi^2 r^2}{L^2}\right) \tag{3}$$

where L is the average size of dots. The Square of the Fourier transform is given by:

$$|C(q)|^2 = \exp\left(-\frac{q^2 L^2}{16\pi^2}\right) \tag{4}$$

The first-order Raman spectrum I(ω) is thus given by:

$$I(\omega) \cong \int \exp\left(-\frac{q^2 L^2}{16\pi^2}\right) \frac{d^3 q}{\left[\omega - \omega(q)\right]^2 + \left(\frac{\Gamma}{2}\right)^2} \tag{5}$$

where q is expressed in units of $\frac{2\pi}{a}$ and a=0.54 nm is the lattice constant of silicon, Γ is the natural line width for c-Si at room temperature (3.5 cm^{-1}) and $\omega(q)$ is the dispersion relation for optical phonons in c-Si which can be taken according to:

$$\omega(q) = \omega_0 - 120\left(\frac{q}{q_0}\right)^2 \tag{6}$$

where $\omega_0 = 520$ cm^{-1} and $q_0 = 2\pi / a_0$.

It was proved that the phonon confinement model is suitable to fit experimental results (Sirleto et al., 2006).

Experimental measurements proving broadening and tuning of Raman spectra have been also performed (Ferrara et al., 2008). Unpolarised Raman spectra have been detected at room temperature in backscattering geometry using a Jobin Yvon Ramanor U-1000 double monochromator, equipped with a microscope Olympus BX40 for micro-Raman sampling and an electrically cooled Hamamatsu R943-02 photomultiplier for photon-counting detection. The excitation source was a Coherent Innova 70 argon ion laser, operating at 514.5 nm wavelength. In order to prevent laser-annealing effects, the average laser power was about 2 mW at the sample surface. Using a 50X objective having long focal distance, the laser beam was focused to a diameter of few microns. Its position on the sample surface was monitored with a video camera. All components of the micro-Raman spectrometer were fixed on a vibration damped optical table.

In Fig.3, the Raman spectra of samples of Si-nc having different size are shown. As expected on the basis of the phonon confinement model (Campbell & Fauchet, 1986; Ritcher at al, 1981), the red shift and the asymmetry of the PS Raman peak increase with decreasing size of Si-nc.

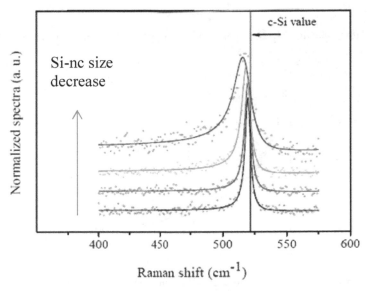

Fig. 3. Raman spectra (circles) of Si-nc samples with different size of nanocrystals.

According to the phonon confinement model, in Fig.4 the peak width of spontaneous Raman emission as a function of nanocrystal size is reported. It is possible to note that the peak width has an inverse dependence on crystal size. Furthermore, considering silicon nanocrystals having crystal size of 2 nm, a significant broadening could be obtained (bandwidth of about 65 cm^{-1}).

Moreover, according to the same model, in Fig.5 the peak shift of spontaneous Raman emission as a function of nanocrystal size is reported. Also the peak shift has an inverse dependence on nanocrystal size. For nanocrystal of about 2 nm a peak shift of about 19 cm-1 can be obtained. Because the width of C-band telecommunication is 146 cm-1, taking into account the broadening and the shift of spontaneous Raman emission, we conclude that more than the half of C-band could be cover considering Si-nc, without implementing the multi-pump scheme.

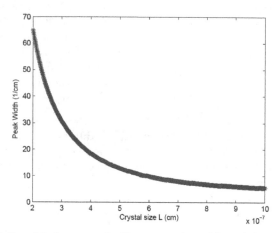

Fig. 4. Calculated relationship between the Raman peak width and the nanocrystal size.

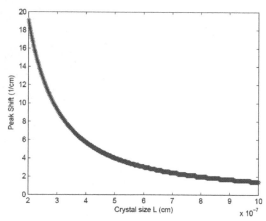

Fig. 5. Calculated relationship between the Raman peak shift and the nanocrystal size.

4. Stimulated Raman Scattering measurements

According with the results discussed above, in this section are reported measurements of stimulated Raman scattering in two different sample: silicon nanocrystals embedded in Si-rich nitride/silicon superlattice structures (SRN/Si-SLs) and silicon nanocomposites embedded in SiO_2.

4.1 Experimental set-up

In Fig. 6, the experimental setup used in order to measure SRS is shown. The pump laser is a CW pump-Raman laser operating at 1427 nm. The probe laser is a tunable external cavity diode laser (1520–1620 nm). The probe beam is split by a Y fiber optic junctions. One of the branches is used in order to monitor probe fluctuations. The other one and the pump laser are combined on a dichroic mirror and subsequently coupled to a long working distance 50X infrared objective in order to be focused onto the sample. The sample is mounted parallel to the path of the incident beam. Estimated coupling losses were about 4 dB. The transmitted signals from the sample are collected by a 20X microscope objective. In order to separate the probe from the pump, a dichroic filter and a longpass filter were used. An optically broadband photodetector (PD) was used to collect the probe signal. The signal from the PD is demodulated by a lock-in amplifier, which is externally referenced to the 180 Hz chopper. Each data point is averaged 1000 times before being acquired. Additionally, four measured values are averaged for each data point. The accuracy of measurement is ±0.1 dB.

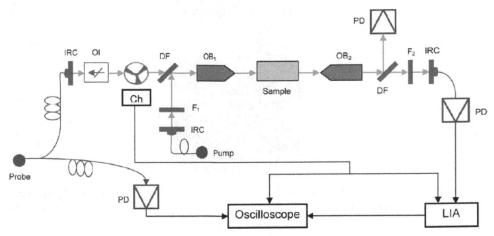

Fig. 6. Experimental setup for SRS measurements: pump-Raman laser; IRC-infra-red collimator; F1-bandpass filter at 1427 nm; probe: ECDLexternal cavity diode laser (tunable); OI-optical insulator; DF-dichroic filter; OB1 (OB2)-microscope objective lens 50X (20X); F2-longpass filter at 1500 nm; Ch-Chopper; PD-optically-broadband photodetector; LIA-Lock-in amplifier. Black lines represent electrical connections and wiring, green lines represent freespace optical beams, and magenta lines represent optical fiber.

4.2 Samples preparation and structural characterization

Raman amplification measurements were performed both on silicon nanocrystals embedded in Si-rich nitride/silicon superlattice structures (SRN/Si-SLs) and on silicon nanocomposities embedded in SiO_2.

SRN/Si-SLs samples were fabricated on Si substrates by radio frequency (RF) magnetron cosputtering from Si, and Si_3N_4 targets. The sputtering of the multilayer structures was performed in a Denton Discovery 18 confocal-target sputtering system, as described elsewhere (Dal Negro et al., 2008). An atomic concentration of 48% Si was measured in the deposited films with energy dispersive x-ray analysis (Oxford ISIS). The multilayer structure has been annealed using a rapid thermal annealing furnace in N_2/H_2 forming gas (5% hydrogen) for 10 min at 800 °C. Thermal annealing resulted in a phase separation process leading to the nucleation of amorphous Si clusters embedded in the Si nitride layers (Dal Negro et al., 2008). Figure 7 shows a TEM bright-field image of our sample in cross-section, taken using a JEOL 2010 TEM operated at 200 KV. The average amorphous Si (a-Si) and SRN layer thicknesses were measured to be 21nm and 26nm, respectively. The structure of the sample consists of 10 SRN layers and 9 amorphous Si (a-Si) layers for a total thickness of 450 nm. A higher magnification of a SRN layer shows the nucleation of amorphous Si nanocrystals marked by the arrows in the inset (panel b). The amorphous Si nanocrystals embedded in the Si nitride layers give rise to strong near-infrared photoluminescence with nanosecond decay dynamics at room temperature, as discussed in details elsewhere (Dal Negro et al., 2006; Dal Negro et al., 2008; Li et al., 2008).

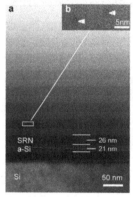

Fig. 7. (a) Cross sectional TEM bright-field micrograph of SRN/Si-SLs consisting of 10 SRN layers and 9 a-Si layers for a total thickness of 450 nm. (b) Higher magnification of a SRN layer, showing amorphous Si nanocrystals marked by arrows.

Silicon nanocomposites sample was obtained by sol-gel technique. Si nanoparticles were obtained by crushing a silicon wafer and reducing their dimensions by a thermal dry oxidation. Afterward, an etching of the oxidized silicon nanoparticles was performed by a solution prepared by hydrofluoric acid and ethanol with a volumetric rate of HF:C_2H_5OH = 1:1. The SiO_2 sol was prepared by mixing the precursor tetraethyl orthosilicate $Si(OC_2H_5)_4$ (TEOS–Sigma–Aldrich) with the solvent (94% denatured ethanol); an homogeneous and continuous film was obtained with a 0.5 M solution. Acidulated water (0.01N HCl:CH_3COOH = 1:1) was added in a hydrolysis ratio HR = 4. Fluorescéine was used as surfactant. Suspension was prepared mixing 5 ml of HF:C_2H_5OH = 1:1 solution with silicon nanoparticles into 20 ml of sol–gel solution. The biggest silicon particles were eliminated filtering the solution by a membrane with pore radius of 0.2 mm. The deposition onto a glass substrate was realised by spin coating. Finally, two thermal treatment, for water and alcohol condensation, were performed. The thin-film so obtained has a thickness of 0.5 mm and its total length is 1.95 cm. The sample was observed using a JEOL JEM 2010F STEM/TEM operated at 200 kV. In Fig.8(a) is shown the EFTEM image in cross view: in white contrast the shape of a crystalline silicon nanoparticle is shown. The STEM analysis in plan view configuration (Fig.8(b)) allowed us to evaluate the mean radius (49 nm) of the silicon dots and also the dot density (1.62×10^8 dots/cm²) (Nicotra et al., 2004; Spinella et al., 2005; Nicotra et al., 2006).

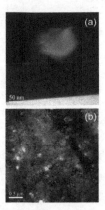

Fig. 8. (a) EFTEM micrograph taken at 16 eV. The white spot is a silicon nanoparticles embedded into the silica matrix. (b) STEM micrograph clearly detects Si nanoparticles, white spots, embedded into the silica matrix.

4.3 SRS Measurements

For both samples we carried out measures of SRS gain. In Fig. 9 and in Fig. 10 (Sirleto et al., 2008; Sirleto et al., 2009; Ferrara et al., 2010), the maxima of the signal wavelength scans are plotted as a function of the effective pump power (including the pass through the filter and objective) for both the SRN/Si-SLs sample and the sample of silicon nanocomposities, respectively.

For SRN/Si-SLs sample, the maximum signal gain obtained was 0.87 dB/cm, while in the case of the sample of silicon nanocomposites the maximum SRS gain obtained was 1.4 dB/cm. Both in Fig.9 and Fig.10, the SRS gain in a float zone high purity and high resistivity bulk silicon is also plotted as a function of the effective pump power. All plots in Fig.9 and in Fig.10 show an approximately linear dependence, as expected for the gain of a Raman amplifier as a function of pump power. As shown in Fig.9, SRN/Si-SLs exhibits a Raman gain significantly greater than bulk silicon. Although the estimation of the gain coefficient g is not straightforward due to the uncertainty in the effective focal volume inside the sample, our data clearly demonstrate a value of g which is about four time larger than the value reported for silicon (Ralston & Chang, 1970) can be obtained in silicon nanostructures. Furthermore, our data prove a threshold power reduction of about 40% in silicon nanocrystals ($P_{th} \approx 150$ mW) with respect to silicon ($P_{th} \approx 250$ mW).

Raman amplifier based on silicon nanocomposites exhibits a SRS gain significantly greater than bulk silicon, as shown in Fig.10. By our data, a preliminary evaluation of approximately a fivefold enhancement of the gain coefficient in Raman amplifier based on silicon nanocomposites with respect to silicon is obtained (Ralston & Chang, 1970). Furthermore our data prove a significant threshold power reduction (about 60%) in silicon nanocomposites ($P_{th} \approx 100$ mW) with respect to silicon ($P_{th} \approx 250$ mW).

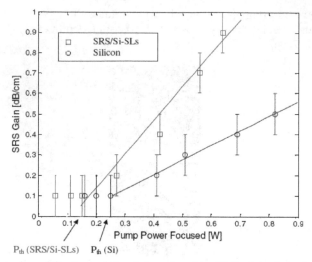

Fig. 9. The SRS-gain (amplification of the stokes signal in dB/cm) is plotted against the effective pump power at the sample surface both for amorphous Si nanoclusters embedded in SRN-Si-SLs (□ red) and for bulk silicon (○ black).

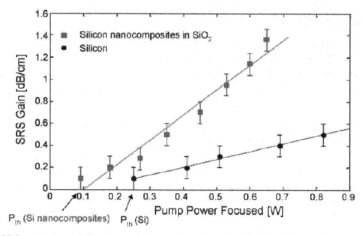

Fig. 10. The SRS gain (amplification of the stokes signal in dB/cm) is plotted against the effective pump power at the sample surface both for silicon nanocomposities in SiO_2 (□ red) and for bulk silicon (○ black).

5. Discussion of results

It is well known that third-order non-linear effects are generally characterized by the non-linear absorption (β) and the non-linear refractive index (γ). The non-linear coefficients, namely β and γ, are described by $a(I) = a_0 + \beta I$ and $n(I) = n_0 + \gamma I$ where a_0 and n_0 stand for the linear absorption and refractive index respectively. The β and γ values are used to evaluate the imaginary (Im $\chi^{(3)}$) and real (Re $\chi^{(3)}$) parts of the third-order non-linear susceptibility, respectively. We note that the real part describes the phenomena related to the intensity dependent index of refraction, while the imaginary part describes two photon absorption and SRS.

A way to enhance the real part of cubic nonlinearities in materials is to artificially 'shrinking' the electrons in regions much shorter than their natural delocalization length in the bulk. In such morphologies, optical resonances will usual appear, resulting from dielectric or quantum confinement, the former prevailing in metal nanocrystals, the latter prevailing in semiconductor nanocrystals. Quantum confinement occurs at a nanometer scale when the electron and hole envelope functions are restricted within a region whose spatial extension is lower than the exciton Bohr radius. This leads to quasi-discrete energy level structures, eventually showing sharp absorption lines. The concentration of the oscillator strength into these discrete levels leads to enhancement of the optical transitions rates and to a size-dependent nonlinear optical susceptibility (Klein, 1996).

Non resonant nonlinearities take place when the light linear absorption is negligible (at frequencies well below the absorption edge). They are related to the anharmonic motion (or virtual excitations) of bound electrons, and are very fast: typical recovery times are of the order of picoseconds. For $\chi^{(3)}$ based nonlinear devices, an enhancement of the real part of the third order nonlinear susceptibility in silicon nanostructure, due to quantum confinement, in the transparency range has been proved (Hernández et al., 2008). It is important to point out that the imaginary part of third order susceptibility of the nanostructured/nanocomposite materials, relating to the Raman gain coefficient, has been never theoretically investigated before. However, an enhancement of imaginary part of third order nonlinear susceptibility is expected.

In order to get Raman laser or amplifier in silicon, the main difficult was due to the presence of TPA, which reduces the efficiency of SRS. Taking advantage of Si-nc optical properties, we explore the possibility to reduce TPA and, at the same time, to enhance SRS.

It is well known that TPA process vanishes for $\hbar\omega < 1/2\ E_g$, $\hbar\omega$ being the photon energy and the band-gap. In Si-nc, due to quantum confinement, an increase of the band-gap with respect to the silicon is obtained. Reducing the dot (wires) dimensions, the band-gap increases. Therefore, considering a suitable size of the dot for which the relation is satisfied, the reduction of TPA is obtained (Lettieri & Maddalena, 2002).

In recent years, there is a strong interest in investigation of Raman scattering in electrons-confined and photons-confined materials. The phenomenon of strong resonant and local enhancement of visible electromagnetic (EM) radiation when incident on the surface of metallic particles and films resulting from surface plasmon resonances continues to attract significant attention for fundamental and applied interests (Kawata et al., 2009). However, the possibility of enhancement of EM radiation from semiconducting and insulating materials, particularly in silicon, is noteworthy for silicon-based optoelectronic applications owing to the potential for monolithically integrating photonic technology and semiconductor electronics.

In the following discussion, we distinguish between Si-nc and silicon nanocomposites, because in the former (silicon nanocrystals) the mean radius of silicon dots was less than in the latter (low dimensional silicon). The different size of the particles corresponds to a fundamental difference between the two experiments reported in this chapter. Concerning stimulated Raman scattering in silicon nanostructures (SRN/Si-SLs), being the particle dimension of about 2 nm, the phonon confinement effect is significant; therefore, we suggest that enhancement explanation has to be sought out in the framework of matter confinement and related to the enhancement of the third order nonlinear susceptibility in small Si clusters (approximately 2nm in diameter) that nucleate in a high density inside silicon-rich nitride materials. However, the structure of the SRN/Si interfaces, the stoichiometric material disorder, and the cluster dimensionality are also important parameters that are expected to significantly influence Raman amplification, a theoretical understanding of their respective roles remains to be established.

Concerning stimulated Raman scattering in silicon nanocomposites in SiO_2, being the diameter of particles dispersed in silica matrix of about 0.1 micron, the enhancement of SRS was due to a photons confinements effect. In order to try to explain why the presence of silicon particles can increase the Raman gain coefficient, we suggest two possible option. The former, it is well known that the nonlinear optical properties of composites material are characterized an "enhancement of local field" (Fischer et al., 1995). Off resonance, the electric field amplitude of an incident laser beam becomes no uniformly distributed between the two constituents of composite and the electric field strength within the more nonlinear constituent will exceed the spatially averaged field strength. Therefore, the effective real part of third order susceptibility of the composite can exceed that of each of its constituents (Fischer et al., 1995). At the same time, strong resonant Raman scattering in dielectric particles is obtained, when the wavelength of an incident field is commensurate with that of an electromagnetic eigenmode of the particle, which depends on its size and on the refractive indices of the particle and the surrounding medium (Murphy & Brueck, 1983). This enhancement can be understood by viewing the particle as a cavity whose dimensions determine whether or not it is in resonance with either the incident or the emitted EM waves or both (i. e. double resonance). The observations of an enhancement of about 100 of the intensity of spontaneous Raman scattering were reported in Murphy & Brueck (1983) from a variety of Si structures having submicrometer dimensions $(x \sim a/\lambda \sim 1)$, where λ is the wavelength and a is a characteristic particle dimension; for example spheres, having diameters of ~0.1micron, were considered. As expected, the enhanced one-phonon spectra reported was a Lorentzian line shape with widths comparable with that of good-quality bulk Si, while the exploited resonances were corresponding to electromagnetic cavity modes with at least one optical wavelength resonating within the particle.

The latter, is related to the optical transport properties of complex photonics structures on the intermediate regime between complete order or disorder. Light waves in disordered materials perform a random walk, which could lead to a multiple scattering process and to a strong localization of dielectric field (Wiersma et al., 2005). In a binary system with components of refractive indices n1 and n2, the efficiency of light scattering depends on how these components are organized in the system, the dimensions of the components, and the refractive index ratio n1/n2=m. In a specific regime, light propagation can be inhibited due to interference and the field intensity in localized regions can be significantly larger than in the surroundings (Schuurmans et al., 1999). As a consequences nonlinear optical properties of disordered material should be enhanced.

In our opinion the localization could play an important role on SRS and the combination of localization and SRS gain could be of particular interest for photonic application, where disordered materials could provide optical amplification via SRS.

6. Conclusions

In this chapter, we experimentally investigate stimulated Raman scattering in amorphous Si nanocrystals and silicon nanocomposite and a number of advantages with respect to silicon were demonstrated.

First, according to phonon confinement model, two significant improvement of Raman approach in silicon quantum dots with respect to silicon were reported: the broadening of spontaneous Raman emission and the tuning of the Stokes shift. Considering silicon quantum dots having crystal size of 2 nm, a significant broadening of about 65 cm^{-1} and a peak shift of about 19 cm^{-1} were obtained. Taking into account such results, more than the half of C-band telecommunication (width = 146 cm^{-1}) could be covered using silicon quantum dots, without implementing the multi pump scheme.

Then, we experimentally demonstrate that amorphous Si nanocrystals and silicon nanocomposite can provide larger Raman gain values and a significant reduction in threshold power with respect to bulk Si devices.

We have two significant consequences related to our results. The first, concerning a fundamental point of view, the broadening of the Raman gain spectra combined with the present observation of enhanced Raman gain lead us to conclude that the traditional tradeoff between gain and bandwidth, valid in bulk materials, could be overcome in low-dimensional materials.

The second, concerning an applicative point of view, the possibility to enhance the Raman gain coefficient and to reduce two-photon absorption, at the same time, in silicon quantum dot was addressed. This means that the main limitation of first silicon Raman lasers could be overcame. Therefore, we quite optimistic believe that our experimental results can open the way to the fabrication of more efficient Raman lasers and amplifiers compatible with Si technology.

7. Acknowledgment

The authors would like to thank Luca Dal Negro, Soumendra N. Basu, Joe Warga and Rui Li for their contributions about the realization and the structural characterization of the sample of silicon nanocrystals embedded in Si-rich nitride/silicon superlattice structures. Moreover, we would like to thank Corrado Spinella and Giuseppe Nicotra for their help about the structural characterization of the sample of silicon nanocomposities embedded in SiO$_2$.

8. References

Agrawal, G. P. (1995). *Nonlinear Fiber Optics*, 2nd ed. New York: Academic.
Boyd, R. W. (2003). Non linear optics, Second edition, 451-476, Academic Press, USA.
Boyraz, O. and Jalali, B. (2004). Demonstration of a silicon Raman laser, *Opt. Express*, Vol. 12, No. 21, pp. 5269–5273.
Campbell, H. and Fauchet, P. M. (1986). The effects of microcrystal size and shape on the one phonon Raman spectra of crystalline semiconductors, *Solid State Commun.*, Vol. 58, 739-741.

Cao, L.; Nabet, B.; Spanier, J. E. (2006). Enhanced Raman scattering from individual semiconductor nanocones and nanowires, *Physical Review Letters*, Vol. 96, no. 15, Article ID 157402, pp. 1–4.

Claps, R.; Dimitropoulos, D.; Han, Y. and Jalali, B. (2002), Observation of Raman emission in silicon waveguides at 1.54 μm, *Opt. Express*, Vol. 10, No. 22, pp. 1305–1313.

Claps, R.; Dimitropoulos, D.; Raghunathan, V.; Han, Y. and Jalali, B. (2003). Observation of stimulated Raman scattering in silicon waveguides, *Opt. Express*, Vol. 11, No. 15, pp. 1731–1739.

Claps, R.; Raghunathan, V.; Dimitropoulos, D. and Jalali, B. (2003). Anti-stokes Raman conversion in silicon waveguides, *Opt. Express*, Vol. 11, No. 22, pp. 2862–2872.

Claps, R.; Raghunathan, V.; Dimitropoulos, D. and Jalali, B. (2004). Influence of nonlinear absorption on Raman amplification in silicon waveguides, *Opt. Express*, Vol. 12, No. 12, pp. 2774–2780.

Dal Negro, L.; Yi, J. H.; Michel, J.; Kimerling, L. C.; Chang, T.-W. F.; Sukhovatkin, V.; Sargent, E. H. (2006). Light emission efficiency and dynamics in silicon-rich silicon nitride films, *Appl. Phys. Lett.*, Vol. 88, pp. 233109.

Dal Negro, L.; Li, R.; Warga, J.; Basu, S. N. (2008). Sensitized Erbium emission from silicon-rich nitride/silicon superlattice structures, *Appl. Phys. Lett.*, Vol. 92, pp. 181105.

Dekker, R.; Usechak, N.; Först, M.; Driessen, A. (2007). Ultrafast nonlinear all-optical processes in silicon-on-insulator waveguides, *J. Phys. D*, Vol. 40, pp. R249-R271,.

Dimitropoulos, D.; Jhaveri, R.; Claps, R.; S.Woo, J. C. and Jalali, B. (2005), Lifetime of photogenerated carriers in silicon-on-insulator rib waveguides, *Appl. Phys. Lett.*, Vol. 86, No. 7, pp. 071115(1)–071115(3).

Dimitropoulos, D.; Fathpour, S.; Jalali, B. (2005). Intensity dependence of the carrier lifetime in silicon Raman lasers and amplifiers, *Appl. Phys. Lett.*, Vol. 87, No. 26, pp. 261108(1)–261108(3).

Espinola, R. L.; Dadap, J. I.; Osgood, R. M. Jr.; McNab, S. J. and Vlasov, Y. (2004). Raman amplification in ultrasmall silicon-on-insulator wire waveguides, *Opt. Express*, Vol. 12, No. 16, pp. 3713–3718.

Fathpour, S.; Boyraz, O.; Dimitropoulos, D.; Jalali, B. (2006). Demonstration of CW Raman gain with zero electrical power dissipation in p-i-n silicon waveguides, presented at the IEEE Conf. Lasers and Electro-Optics (CLEO), Long Beach, CA, 2006, Paper CMK3.

Ferrara, M. A.; Donato, M. G.; Sirleto, L.; Messina, G.; Santangelo, S.; Rendina, I. (2008). Study of strain and wetting phenomena in porous silicon by Raman scattering, *Journal of Raman Spectroscopy*, Vol. 39, pp. 199-204.

Ferrara, M. A.; Sirleto, L.; Nicotra, G.; Spinella, C.; Rendina, I. (2010). Enhanced gain coefficient in Raman amplifier based on silicon nanocomposites, *Photon Nanostruct: Fundam. Appl.*, http://dx.doi.org/10.1016/j.photonics.2010.07.007

Freeouf, J. L.and Liu, S. T. (1995). Minority carrier lifetime results for SOI wafers, in *Proc. IEEE Int. SOI Conf.*, Oct. 1995, pp. 74–75.

Fischer, G. L.; Boyd, R. W.; Gehr, R. J.; Jenekhe, S. A.; Osaheni, J. A.; Sipe, J. E.; Weller-Brophy, L. A. (1995). Enhanced nonlinear optical response of composite materials, *Phys. Rev. Lett.*, Vol. 74, pp. 1871.

Hart, T. R.; Aggarwal, R. L. and Lax, B. (1970). Temperature dependence of Raman scattering in silicon, *Phys. Rev. B, Condens. Matter*, Vol. 1, No. 2, pp. 638–642.

Hernández; Pellegrino, P.; Martínez, A.; Lebour, Y.; Garrido, B.; Spano, R.; Cazzanelli, M.; Daldosso, N.; Pavesi, L.; Jordana, E.; Fedeli, J. M. (2008). Linear and nonlinear optical properties of Si nanocrystals in SiO_2 deposited by plasma-enhanced chemical-vapor deposition, *Journal of Applied Physics*, Vol. 103, pp. 064309.

Islam, M.N. (2002). Raman amplifiers for telecommunications, *IEEE J. Selected Top. Quantum Electron.*, Vol. 8, pp. 548–559.

Jalali, B.; Raghunathan, V.; Dimitropoulos, D.; Boyraz, O. (2006). Raman-based silicon photonics, *IEEE J. Sel. Top. Quantum Electron.*, Vol. 12, pp. 412-421.

Jian Wu; Awnish K. Gupta; Humberto R. Gutierres, Eklund, P. C. (2009). Cavity-enhanced stimulated Raman scattering from short GaP nanowires, *Nano Lett.*, Vol. 9, pp. 3252-3257.

Kawata, S.; Inouye, Y.; Verma, P. (2009). Plasmonics for near-field nano-imaging and superlensing, *Nature Photonics*, Vol. 3, no. 7, pp. 388–394.

Klein, L. C. (1996). Nanomaterials: Synthesis, Properties and Application, A. S. Edelstein and R. C. Cammarata, Institute of Physics Bristol-UK.

Lettieri, S. and Maddalena, P. (2002). Nonresonant Kerr effect in microporous silicon: Nonbulk dispersive behavior of below band gap $\chi_{(3)}(\omega)$, *J. Appl. Phys.*, Vol. 91, 5564.

Li, R.; Schneck, J. R.; Warga, J.; Ziegler, L. D.; Dal Negro, L. (2008). Carrier dynamics and erbium sensitization in silicon-rich nitride nanocrystals, *Appl. Phys. Lett.*, Vol. 93, pp. 091119.

Liang, T. K. and Tsang, H. K. (2004). Role of free carriers from two-photon absorption in Raman amplification in silicon-on-insulator waveguides, *Appl. Phys. Lett.*, Vol. 84, No. 15, pp. 2745–2747.

Liu, A.; Rong, H.; Paniccia, M.; Cohen, O.; Hak, D. (2004). Net optical gain in a low loss silicon-on-insulator waveguide by stimulated Raman scattering, *Opt. Express*, Vol. 12, no. 18, pp. 4261–4267.

Mendicino, M. A. (1998). Comparison of properties of available SOI materials, in *Properties of Crystalline Silicon*, Hull, R.; Ed. London, U.K.: Inst. Eng. Technol., 1998 , ch. 18.1, pp. 992–1001.

Mori, A.; Masuda, H.; Shikano, K.; Shimizu, M. (2003). Ultra-wide-band tellurite-based fiber Raman amplifier, *J. Lightwave Technol.*, Vol. 21, pp. 1300.

Murphy, D. V.; Brueck, S. R. J. (1983). Enhanced Raman scattering from silicon microstructures, *Optics Letters*, Vol. 8, pp. 494-496.

Nature News Editorial. *First silicon laser pulses with life*. Available: from http://www.nature.com/news/2004/041025/full/041025-10.html

Nicotra, G.; Puglisi, R.A.; Lombardo, S.; Spinella, C.; Vulpio, M.; Ammendola, G.; Bileci, M.; Gerardi, C. (2004). Nucleation kinetics of Si quantum dots on SiO_2, *J. Appl. Phys.*, Vol. 88, pp. 2049.

Nicotra, G.; Hui-Ting Chou, Henning Stahlberg, N.D. ; Browning, (2006). Avoiding charge induced drift in vitrified biological specimens through scanning transmission electron microscopy, *Microsc. Microanal.* Vol. 12, pp. 246–247.

OSA Press Room Editorial. *Advancing the science of light*. Available from: http://www.osa.org/news/pressroom/release/11.2002/jalali.aspx

Pavesi, L.; Lockwood, D.J. (2004). *Silicon Photonics*, Springer-Verlag, Berlin 2004.

Ralston, J. M. and Chang, R. K. (1970). Spontaneous-Raman-scattering efficiency and stimulated scattering in silicon, *Phys. Rev. B, Condens. Matter*, Vol. 2, No. 6, pp. 1858–1862.

Ritcher, H.; Wang, Z. P.; Ley, L. (1981). The one phonon Raman spectrum in microcrystalline silicon, *Solid State Commun.*, Vol. 39, pp. 625-629.

Rivero, C.; Richardson, K.; Stegeman, R.; Stegeman, G.; Cardinal, T.; Fargin, E.; Couzi, M.; Rodriguez, V. (2004). Quantifying Raman gain coefficients in tellurite glasses, *J. Non-Cryst. Solids*, Vol. 345–346, pp. 396-401,.

Rong, H.; Liu, A.; Jones, R.; Cohen, O.; Hak, D.; Nicolaescu, R.; Fang, A. and Paniccia, M. (2005). An all-silicon Raman laser, *Nature*, Vol. 433, No. 7023, pp. 292–294.

Rong, H.; Jones, R.; Liu, A.; Cohen, O.; Hak, D.; Fang, A. and Pannicia, M. (2005). A continuous-wave Raman silicon laser, *Nature*, Vol. 433, No. 7027, pp. 725–728.

Rong, H.; Kuo, Y. H.; Xu, S.; Liu, A.; Jones, R.; Paniccia, M. (2006). Monolithic integrated Raman silicon laser, *Opt. Express*, Vol. 14, No. 15, pp. 6705-6712.

Russell, J. P. (1965). Raman scattering in silicon, *Appl. Phys. Lett.*, Vol. 6, No. 11, pp. 223–224.

Shen, Y.R.; Bloembergen, N. (1965). Theory of stimulated Brillouin and Raman scattering, *Phys. Rev.*, Vol. 137, pp. A1787.

Schuurmans, F. J. P.; Vanmaekelbergh, D.; van de Lagemaat, J.; Lagendijk, A. (1999). Strongly Photonic Macroporous GaP Networks, *Science*, Vol. 284, pp. 141-143.

Sirleto, L.; Ferrara, M. A.; Rendina, I.; Jalali, B. (2006). Broadening and tuning of spontaneous Raman emission in porous silicon at 1.5 micron, *Appl. Phys. Lett.*, Vol. 88, pp. 211105.

Sirleto, L.; Ferrara, M. A.; Rendina, I.; Basu, S.N.; Warga, J.; Li, R.; Dal Negro, L (2008). Enhanced stimulated Raman scattering in silicon nanocrystals embedded in silicon-rich nitride/silicon superlattice structures, *Applied Physics Letters*, Vol. 93, pp. 251104.

Sirleto, L.; Ferrara, M.A.; Nicotra, G; Spinella, C; Rendina, I. (2009). Observation of stimulated Raman scattering in silicon nanocomposites, *Appl. Phys. Lett.*, Vol. 94, pp. 221106.

Spillane, S. M.; Kippenberg, T. J.; Vahala, K. J. (2002). Ultralow-threshold Raman laser using a spherical dielectric microcavity, *Nature*, Vol. 415, pp. 621-623.

Spinella, C.; Bongiorno, C.; Nicotra, G.; Rimini, E.; Muscarà, A.; Coffa, S. (2005). Quantitative determination of the clustered silicon concentration in substoichiometric silicon oxide layer, *Appl. Phys. Lett.*, Vol. 87, pp. 044102.

Soref, R. (2006). The Past, Present, and Future of Silicon Photonics, *IEEE Journal of Selected Topics in Quantum Electronics*, Vol. 12, No. 6, pp. 1678-1687.

Temple, P. A. and Hathaway, C. E. (1973). Multiphonon Raman spectrum of silicon, *Phys. Rev. B, Condens. Matter*, Vol. 7, No. 8, pp. 3685-3697.

Wiersma, D. S.; Sapienza, R.; Mujumdar, S.; Colocci, M.; Ghulinyan, M.; Pavesi, L. (2005). Optics of nanostructured dielectrics, *J. Opt. A, Pure Appl. Opt.*, Vol. 7, pp. S190.

Zhang, B. P.; Shimazaki, K.; Shiokawa, T.; Suzuki, M.; Ishibashi, K.; Saito, R. (2006). Stimulated Raman scattering from individual single-wall carbon nanotubes, *Appl. Phys. Lett.*, Vol. 88, pp. 241101.

Part 2

Unique Nonlinear Optics Schemes and Network

4

Nonlinear Ellipsometry by Second Harmonic Generation

Fabio Antonio Bovino[1], Maria Cristina Larciprete[2], Concita Sibilia[2]
Maurizio Giardina[1], G. Váró[3] and C. Gergely[4]
[1]*Quantum Optics Lab Selex-Sistemi Integrati, Genova, Italy*
[2]*Department of Basic and Applied Sciences in Engineering, Sapienza University, Rome,*
[3]*Institute of Biophysics, Biological Research Center,*
Hungarian Academy of Sciences, Szeged,
[4]*Montpellier University, Charles Coulomb Laboratory UMR 5221, Montpellier,*
[1,2]*Italy*
[3]*Hungary*
[4]*France*

1. Introduction

Among the different nonlinear optical processes, second harmonic generation (SHG) is one of the most investigated. Briefly, polarization in a dielectric material can be expanded in terms of applied electric field. Second harmonic generation corresponds to an optical process of coherent radiation from electric-dipoles forming in the nonlinear optical material. In particular, SHG is related to the second term of the polarization expansion, thus it can be obtained only in materials which are noncentrosymmetric i.e. posses no centre of inversion symmetry. From the experimental point of view, the frequency of the incoming – fundamental- beam, ω, is doubled by the second order optical susceptibility $\chi_{ijk}^{(2)}$ of the material. The SHG processes, along with the structure of the nonlinear optical tensor, $\chi_{ijk}^{(2)}$, are strongly dependent on the crystalline structure of the material, thus by choosing the appropriate polarization state for the fundamental beam, different amplitude and polarization state of the nonlinear optical response can be selectively addressed.

As a consequence, several experimental techniques have been developed, for the determination of the different non-zero components of the third rank tensor $\chi_{ijk}^{(2)}$, with reference to a well-characterized sample. The Maker fringes technique (Maker et al, 1962), which is based on the investigation of oscillations of the SH intensity by changing the crystal thickness, has been without doubt the most employed. Briefly, this technique consists in measuring the SH signal transmitted trough the nonlinear crystal as a function of the fundamental beam incidence angle, which is continuously varied by placing the sample onto a rotation stage. The polarization states of both fundamental and generated beams are selected by rotating a half-wave plate (polarizer) and a linear polarizer (analyzer), respectively. On a reference line, a small fraction of the fundamental beam is usually sent onto a reference crystal, which is hold at a fixed incidence angle, in order to minimize the influence of laser energy fluctuations. On the measurement line, the second harmonic signal is detected with a photomultiplier, while interference and dichroic filters are used to suppress the fundamental beam.

Second harmonic signal can also be generated at a surface, being itself responsible for a symmetry break (Bloembergen et al, 1968). When looking for surface contributions to the SH signal, rather than bulk contribution, the reflective second harmonic generation (RSHG) technique has to be employed. This technique involves the detection of the reflected SH signal, at a fixed incidence angle of the pump beam, while sample is rotated along its surface normal. Specifically, depending on the form of the bulk $\chi_{ijk}^{(2)}$ tensor, there may be some particular combinations of the polarization states of fundamental and generated beams may inhibit the bulk induced SHG. As a consequence, any signal measured in these polarizations combinations would be ascribed to surface effects.

The noncollinear scheme of SHG experiments was firstly introduced by Muenchausen (Muenchausen et al, 1987) and Provencher (Provencher et al, 1993) and, since then, it was exploited more recently by different authors. It presents some advantages, with respect to conventional collinear SHG, as a reduced coherence length (Faccio et al, 2000) as well as the possibility to distinguish between bulk and surface responses (Cattaneo & Kauranen, 2005) thus this technique represents a promising tool for surface and thin-film characterization (Cattaneo & Kauranen, 2003).

Very recently, we developed a method, based on the noncollinear scheme of SHG, to evaluate the non-zero elements of the nonlinear optical susceptibility. At a fixed incidence angle, the generated noncollinear SH signal is investigated while continuously varying the polarization state of *both* fundamental beams. The obtained experimental results show the peculiarity of the nonlinear optical response associated with the noncollinear excitation, and can be fully explained using the expression for the effective second order optical nonlinearity in noncollinear scheme. The resulting *polarization chart*, recorded for a given polarization state of the SH signal, shows pattern which is characteristic of the investigated crystalline structure. It offers the possibility to evaluate the ratio between the different non-zero elements of the nonlinear optical tensor. Moreover, if the measurements are performed with reference to a well-characterized sample, i.e. a nonlinear optical crystal as quartz or KDP, this method allows the evaluation of the absolute values of the non-zero terms of the nonlinear optical tensor, without requiring sample rotation. As a consequence, this technique turns out to be particularly appropriate for those experimental conditions where the generated SH signal can be strongly affected by sample rotation angle. For instance, if a sample is some coherence lengths thick, as the optical path length is changed by rotation, the SH signal strongly oscillates with increasing incidence angle (Jerphagnon & Kurtz, 1970) according to Maker-fringes pattern, thus a high angle resolution would be required. When using short laser pulses, whose bandwidth is comparable or lower than sample thickness, as the incidence angle is modified the nonlinear interaction may involve different part of the sample and, eventually, surface contributions. For nano-patterned samples, finally, a rotation would imply differences into sample surface interested by the pump spot size. With respect to the mentioned examples, the method of polarization scan simplifies the characterization of the nonlinear optical tensor elements without varying the experimental conditions, and turns out to be a sort of *nonlinear ellissometry*.

In what follows, we will describe in details some applications that we recently developed, where the polarization mapping is employed for the characterization of some nonlinear optical materials as gallium nitride (GaN), zinc oxide (ZnO) and, more specifically to Bacteriorhodopsin films.

2. Evaluation of the non-zero elements of the $\chi^{(2)}$ tensor components

As far as noncollinear SHG is concerned, as in our recent works, the number of experimental parameters which can be combined, so to determine the polarization and amplitude of the SH signal, is increased. As a matter of fact, the two pump beams, tuned at $\omega_1 = \omega_2$, having different incidence angles, α_1 an α_2, and polarization state, ϕ_1 and ϕ_2, cooperate in the determination, and thus in the excitation, of the nonlinear optical polarization, $P^{(2)} = \chi^{(2)} : E_1(\omega_1) E_2(\omega_2)$.

We successfully tested this kind of nonlinear ellisometry onto a Gallium nitride slab, 302 nm thick, grown by metal-organic chemical vapour deposition (MOCVD) onto (0001) c-plane Al2O3 substrates (Potì et al, 2006). GaN presents a wurtzite crystal structure without centre of inversion, thus leading to efficient second order nonlinear effects (Miragliotta et al, 1993). In addition, the wide transparency range, which extends form IR to the near UV, make this material extremely appealing from nonlinear optical point of view.

We employed the output of a mode-locked femtosecond Ti:Sapphire laser system tuned at $\lambda = 830$ nm (76 MHz repetition rate, 130 fs pulse width), which was split into two beams of about the same intensity. The polarization state of both beams (ϕ_1 and ϕ_2) was varied with two identical half wave plates, automatically rotating, that were carefully checked not to give nonlinear contribution. Two collimating lenses, 150 mm focal length, were placed thereafter, while the sample was placed onto a motorized combined translation and rotation stage which allowed the variation of the rotation angle, α, with a resolution of 0.5 degrees. The temporal overlap of the incident pulses was automatically controlled with an external delay line. Several details of the experimental scheme are given in Fig. 1.

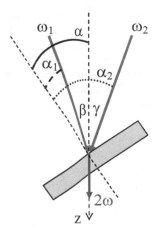

Fig. 1. Sketch of the noncollinear scheme adopted for second harmonic generation measurements. For a fixed sample rotation angle α, measured with respect to the z-axis, the corresponding incidence angles of the two pump beams result to be $\alpha_1 = \alpha - \beta$ and $\alpha_2 = \alpha - \gamma$, respectively.

In the experiments reported in (Larciprete et al, 2009) the sample rotation angle was fixed to $\alpha=35°$, while the pump beams were sent to intersect in the focus region with the angles $\beta=9°$ and $\gamma= -9°$, i.e. the corresponding incidence angles of the two pump beams onto the sample result to be $\alpha_1=\alpha-\beta$ and $\alpha_2=\alpha-\gamma$, respectively.

GaN crystal structure, i.e. wurtzite, is characteristic of III-V nitrides and presents the noncentrosymmetric point group symmetry 6mm with a hexagonal primary cell. The only nonvanishing second order susceptibility tensor elements (J.Chen, Z.H.Levine, J.W.Wilkins, Appl. Phys. Lett. 66, pp. 1129-1131 (1995)) are $\chi_{311}^{(2)} = \chi_{322}^{(2)}$, $\chi_{333}^{(2)}$, and $\chi_{113}^{(2)} = \chi_{131}^{(2)} = \chi_{223}^{(2)} = \chi_{232}^{(2)}$, which correspond to $\chi_{15}^{(2)}=\chi_{24}^{(2)}$, $\chi_{31}^{(2)}=\chi_{32}^{(2)}$ and $\chi_{33}^{(2)}$, referring to the piezoelectric contraction, or equivalently, being $\tilde{d}_{ij} = \frac{1}{2}\tilde{\chi}_{ij}^{(2)}$ the second order nonlinear optical tensor can be written as follows:

$$\tilde{d} = \begin{pmatrix} 0 & 0 & 0 & 0 & d_{15} & 0 \\ 0 & 0 & 0 & d_{24} & 0 & 0 \\ d_{31} & d_{32} & d_{33} & 0 & 0 & 0 \end{pmatrix} \tag{1}$$

The five non-zero terms further reduce to three independent coefficients in wavelength regimes where it is possible to take advantage of Kleinmann symmetry rules, i.e. $d_{15}=d_{24}=d_{31}=d_{32}$ and $d_{33}= -2 \cdot d_{31}= - 2 \cdot d_{15}$.

Given the tensor (1), by selecting the appropriate polarization state for the two fundamental beams, it is possible to address the different non-zero components of $d_{ij}^{(2)}$ and, consequently, to get different polarization state for the generated signal.

The full expression of the SH power, $W_{\omega1+\omega2}$, as a function of sample rotation angle α, is given by:

$$W_{\omega1+\omega2}(\alpha) = \left(\frac{512\pi^3}{A_1 A_2}\right)(t_{\omega1})^2 \cdot (t_{\omega2})^2 \cdot T_{\omega1+\omega1} \cdot W_{\omega1} \cdot W_{\omega2} \frac{\sin^2(\Psi_{SHG}(\alpha))}{\left[n_{\omega1} \cdot n_{\omega2} - n^2_{\omega1+\omega2}\right]^2}(d_{eff}(\alpha))^2 , \tag{2}$$

where A_1 and A_2 are the fundamental beams transverse areas onto sample surface, retrieved from the main beam area (A), $W_{\omega1}$ and $W_{\omega2}$ are the power of the incident fundamental beams. The Fresnel transmission coefficients for the two fundamental fields at the input interfaces are $t_{\omega1}(\alpha_1,\phi_1)$ and $t_{\omega2}(\alpha_2,\phi_2)$, while $T_{\omega1+\omega2}(\alpha,\phi)$ is the Fresnel transmission coefficient for the SH power at the output interface. As far as material optical birefringence is concerned, Fresnel coefficients and refractive indices of both fundamental and generated beams, i.e. $n_{\omega1}(\alpha'_1,\phi_1)$, $n_{\omega2}(\alpha'_2,\phi_2)$ and $n_{\omega1+\omega2}(\alpha'_{\omega1+\omega2},\phi)$ are dependent on the propagation angle and polarization state of the respective beam.

Finally, Ψ_{SHG} is the phase factor given by:

$$\Psi_{SHG} = \left(\frac{\pi L}{2}\right)\left(\frac{2}{\lambda}\right)\left[n_{\omega1} \cdot \cos(\alpha'_{\omega1}) + n_{\omega2} \cdot \cos(\alpha'_{\omega2}) - 2n_{\omega1+\omega2} \cdot \cos(\alpha'_{\omega1+\omega2})\right] , \tag{3}$$

where L is sample thickness.

The analytical expression of the effective nonlinear susceptibility, $d_{eff}(\alpha)$ can be rather complicated, being dependent on the tensor components, the polarization state of the three electric fields and, of course, on the fundamental beams incidence angles. However, for point group symmetry 6mm, as in the case of GaN, the final expressions for $d_{eff}(\alpha)$, as a function of polarization angle of the two pumps, becomes:

$$d_{eff}^{\hat{s}} = -d_{15}\left[\sin(\phi_1)\cos(\phi_2)\sin(\alpha'_2)+\cos(\phi_1)\sin(\phi_2)\sin(\alpha'_1)\right]$$

$$\begin{aligned}d_{eff}^{\hat{p}} = &-d_{24}\cos(\alpha)\left[\cos(\alpha'_1)\sin(\alpha'_2)+\sin(\alpha'_1)\cos(\alpha'_2)\right]\cos(\phi_1)\cos(\phi_2)-\\ &-\sin(\alpha)\left\{d_{31}\sin(\varphi_1)\sin(\varphi_2)+\cos(\phi_1)\cos(\phi_2)\left[d_{32}\cos(\alpha'_1)\cos(\alpha'_2)+d_{33}\sin(\alpha'_1)\sin(\alpha'_2)\right]\right\}\end{aligned} \quad (4)$$

where the apex stands for the polarization state of the generated beam and α'_1, α'_2 are the internal propagation angles of the two pump beams inside the sample. Equations (4) quite completely and exhaustively describe the interaction of two incident pump beams linearly polarized with a noncentrosymmetric material presenting GaN crystalline structure. For any other generic polarization angle of the SH beam, ϕ, the $d_{eff}(\alpha)$ results in a combination of terms given by the Eeq.(4): $d_{eff}^{\phi} = \sin(\phi)d_{eff}^{\hat{s}} + \cos(\phi)d_{eff}^{\hat{p}}$.

Following these considerations, we measured the generated signal as a function of the polarization state of both pump beams, at three different sample rotation angles, i.e. for α=35 , 9 nd 1 degrees. The two half-wave plates were systematically rotated, in the range -180 -+180 degrees for the first pump beam (ϕ_1) and 0 -180 degrees for the second pump beam (ϕ_2).

We show the obtained measurements in Fig.2.a and Fig.2.b for the two different polarization state of the analyzer, namely $\hat{p}$ ϕ_1=0°, and $\hat{s}$, ϕ_2=90°, respectively. Considering the $\hat{p}$ - polarized SH signal (Fig.2.a) it can be seen that the absolute maxima are achievable when both pumps are $\hat{p}$ -polarized, i.e. when ϕ_1 and ϕ_2 are both 0° or 180°, while relative maxima occur when both pumps are $\hat{s}$ -polarized, i.e. when polarization angles of both pumps are set to ± 90°. Conversely, when the two pump beams have crossed polarization, i.e when ϕ_1=0° and ϕ_2=90° and viceversa, the nonlinear optical tensor of GaN do not allow second harmonic signal which is $\hat{p}$ -polarized thus the corresponding measurements go to zero.

A fairly different behavior is observable, when the analyzer is set to $\hat{s}$ -polarization, i.e. ϕ=90° (see Fig.2.b). In this case, the maxima generally occur when the two pump beams have crossed polarization, but since this condition is no more symmetrical for positive and negative rotation angles, the resulting surface plots present some variation at different rotation angles. When α=35° (Fig.2.b), the absolute maxima take place when the first pump is $\hat{s}$ -polarized and the second pump is $\hat{p}$ -polarized, i.e. ϕ_1= ±90° and ϕ_2 is equal to either 0° or 180°. Relative maxima occur in the reverse situation, when the first pump is $\hat{p}$ -polarized, ϕ_1=0° or ±180°, and the second pump $\hat{s}$ -polarized, ϕ_2=90°. Finally, if the two pumps are equally polarized, either $\hat{s}$ or $\hat{p}$, the generation of $\hat{s}$ -polarized signal is not allowed.

The calculated polarization charts, reported in Fig.3.a and Fig.3.b, were retrieved from Equations (4) by assuming the Kleinmann symmetry rules for the nonlinear optical tensor elements.

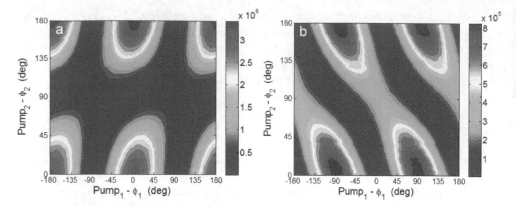

Fig. 2. Noncollinear second harmonic signal experimentally measured as a function of the polarization state of the first pump beam (ϕ_1) and the second pump beam (ϕ_2). Sample rotation angle was fixed to α=35° . The polarization state of the analyzer is set to (a) $\hat{p}$ i.e. ϕ=0° and (b) $\hat{s}$, i.e. ϕ=90°.

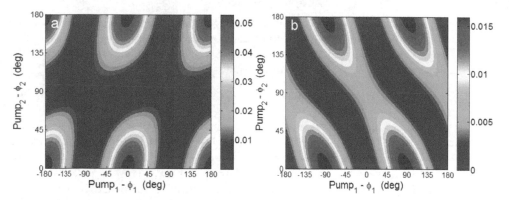

Fig. 3. Noncollinear second harmonic signal theoretically calculated as a function of the polarization state of the first pump beam (ϕ_1) and the second pump beam (ϕ_2). Sample rotation angle was fixed to α=35°. The polarization state of the analyzer is set to (a) $\hat{p}$ i.e. ϕ=0° and (b) $\hat{s}$, i.e. ϕ=90°.

The perfect matching between the experimental and theoretical charts verify the rightness of the symmetry assumption. Assuming a different relationship between the coefficients d_{15}, d_{31} and d_{33} would in fact lead to evident changes in the polarization charts.

In order to evaluate the effect of sample rotation angle, we performed further experimental measurements at different sample rotation angles. The experimental plots obtained for α=1° and 9° are shown in Fig.4 and Fig.5, respectively.

The polarization charts of the noncollinear SH signal generated in $\hat{p}$ polarization (see Fig.4.a and Fig.5.a) display a similar symmetry at all the sample rotation angles, while amplitude is decreasing with decreasing rotation angle.

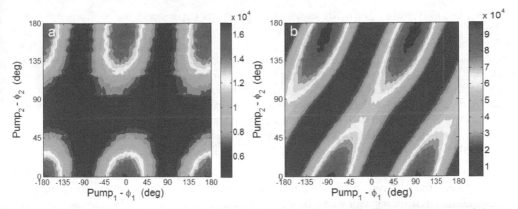

Fig. 4. Noncollinear second harmonic signal experimentally measured as a function of the polarization state of the first pump beam (ϕ_1) and the second pump beam (ϕ_2). Sample rotation angle was fixed to α=1°. The polarization state of the analyzer is set to (a) $\hat{p}$ i.e. ϕ=0° and (b) $\hat{s}$, i.e. ϕ=90° .

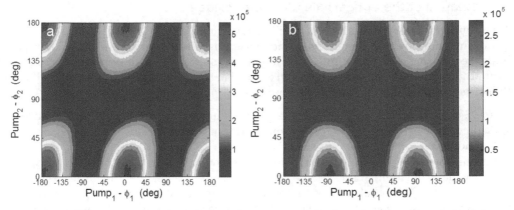

Fig. 5. Noncollinear second harmonic signal experimentally measured as a function of the polarization state of the first pump beam (ϕ_1) and the second pump beam (ϕ_2). Sample rotation angle was fixed to α=9°. The polarization state of the analyzer is set to (a) $\hat{p}$ i.e. ϕ=0° and (b) $\hat{s}$, i.e. ϕ=90°.

On the other side, the $\hat{s}$-polarized SH signal (see Fig.4.b and Fi.5.b), according with the theoretical model, result in a modified trend of both the Fresnell coefficients and the effective nonlinearity as a function of ϕ_1 and ϕ_2. As a consequence, when the rotation angle α is set to 1° (Fig.4.b) we found that the plots appear to be reversed, with respect to α =35°.

Curiously, when α =9° the same conditions hold for the absolute maxima and for the zero signal, while the relative maxima disappeared. This unusual behavior can be explained considering that fixing the sample rotation angle to 9°, i.e. fixing α=β, corresponds to a situation such that the first pump beam is normally incident onto the sample. For an anisotropic uniaxial crystal with the optical axis perpendicular to sample surface, as the investigated GaN film, a normally incident wave always experiences the ordinary refractive index, whatever its polarization angle. Thus, from the refractive index point of view, the polarization state of the first pump beam always corresponds to the case of $\hat{s}$ -polarization. As a consequence, the condition to get the relative maxima, i.e. the first pump $\hat{p}$ -polarized and the second pump $\hat{s}$ -polarized, is never fulfilled, since it is replaced with the combination of two pumps both having $\hat{s}$ -polarization and the SH generation of $\hat{s}$ - polarized signal is prohibited. As we will show in the next section, this experimental configuration, i.e. one of the pump is normally incident onto the sample, is particularly suited to put evidence a tilt in the optical axis, since it would result in a modified pattern of the $\hat{s}$ polarized signal.

Finally, we have shown that the *polarization charts* offer all the information to evaluate the ratio between the different non-zero elements of the nonlinear optical tensor, thus verifying if Kleinman's symmetry rules can be applied to a given material. The method we have described is an extension of Maker fringes technique to the noncollinear case and represents a useful tool to characterize the non-zero terms of the nonlinear optical tensor without varying relevant experimental conditions as incidence angles.

3. Evaluation of the optical axis tilt of Zinc oxide films

We applied the noncollinear nonlinear ellissometry to ZnO films grown by dual ion beam sputtering and show that the proposed nonlinear ellissometry is an useful tool to put into evidence a tilt angle of the optical axis of a nonlinear optical film with respect to the surface normal, for any material whose symmetry class implies an orientation of the optical axis almost perpendicular to the surface (Bovino et al, 2009).

Zinc Oxide was chosen for the large energy gap value (Eg = 3.37 eV) and high nonlinear optical coefficients, of both second and third order, it offers (Blachnik et al, 1999). Second order nonlinear optical response has been shown in ZnO films grown by different techniques implying both high deposition temperature (as reactive sputtering, spray pyrolysis, laser ablation) and low deposition temperature (as laser deposition, and dual ion beam sputtering). Generally, the reduced deposition temperature results in polycrystalline films, where the average orientation of crystalline grains, along with the resulting optical axis, can be tilted with respect to the ideal crystal, i.e. normal to sample surface.

Zinc oxide films, 400 nm thick, were deposited by means of a dual ion beam sputtering system onto silica substrates. Preliminary X-ray diffraction investigation performed on the obtained films indicate that the films are polycrystalline with the c-axis preferentially oriented about the surface normal (Weißenrieder & Muller, 1997).

As well as for GaN, ZnO crystalline structure belongs to the noncentrosymmetric point group symmetry 6mm with a hexagonal primary cell, thus the non-zero components are the same, i.e. $d_{15}=d_{24}=d_{31}=d_{32}$ and $d_{33}= -2 \cdot d_{31}= - 2 \cdot d_{15}$, under Kleinmann's approximation. However, it must be pointed out that this assumption holds only if the optical axis is normal to the sample surface. If, on the other hand, the optical axis is somehow tilted, with respect to the surface normal, a rotation must be introduced into the expression of the nonlinear optical tensor, that in turns results into the introduction of other nonvanishing terms in the effective nonlinear susceptibility.

The analytical expression of the effective susceptibility, $d_{eff}(\alpha)$, for the ZnO crystalline structure, considering four combination of polarization states of the two pump beams, $\hat{p}_{\omega 1} - \hat{p}_{\omega 2}$, $\hat{s}_{\omega 1} - \hat{s}_{\omega 2}$, $\hat{p}_{\omega 1} - \hat{s}_{\omega 2}$ and $\hat{s}_{\omega 1} - \hat{p}_{\omega 2}$, four different expressions for $d_{eff}(\alpha)$ are allowed, depending on the SH polarization state, i.e. either $\hat{p}$ or $\hat{s}$:

$$d_{eff}^{pp \to p} = -\cos(\alpha_{2\omega})d_{24}\left[\cos(\alpha_1')\sin(\alpha_2') + \cos(\alpha_2')\sin(\alpha_1')\right] +$$
$$+ \sin(\alpha_{2\omega})\left[-d_{32}\cos(\alpha_1')\cos(\alpha_2') - d_{33}\sin(\alpha_2')\sin(\alpha_1')\right]$$
$$d_{eff}^{ss \to p} = -\sin(\alpha_{2\omega})d_{31} \tag{5}$$
$$d_{eff}^{ps \to s} = -d_{15}\sin(\alpha_1')$$
$$d_{eff}^{sp \to s} = -d_{15}\sin(\alpha_2')$$

Where α_1', α_2' are the internal propagation angles of the two pump beams inside the sample, and $\alpha_{2\omega}$ is the angle of emission of the SHG inside the crystal.

Referring to Fig.1, the two pump beams were sent to intersect in the focus region with the angles $\beta = 9°$ and $\gamma = -9°$, while α was fixed to 9°. As a matter of fact, the fundamental beam 1 was normally incident onto the sample. The experimental measurements were obtained by rotating the two half-wave plates, in the range -180° -+180° for pump beam 1 and 0° -180° for pump beam 2.

The experimental plots, obtained when the analyzer was set to $\hat{s}$ -polarization, are shown in Figure 6.a. As we already mentioned in the previous section, the maxima of SH signal should occur when the two pump beams have crossed polarization. However, in this particular condition, the absolute maxima still require $\hat{s}$ -polarization for pump 1 and $\hat{p}$ -polarization for pump 2 (i.e. $\phi_1 = \pm 90$ and ϕ_2 equal to either 0 or 180), whereas the relative maxima totally disappeared. In this configuration, in fact, the pump beam 1 is normally incident onto the sample (see Figure 6.b) thus it is always $\hat{s}$ -polarized, i.e. the condition to get a relative maximum ($\hat{p}$ -polarization for pump 1 and $\hat{s}$ -polarization for pump 2) vanishes. What is even more interesting, we found out that the experimental configuration where one of the pump beams is normally incident onto the sample, is particularly sensitive to the orientation of the optical axis.

The experimental curves were fully reconstructed using the expression for the effective second order optical nonlinearity in noncollinear scheme, assuming the Kleinmann symmetry rules. Dispersion of both the ordinary and extraordinary refractive indices of ZnO are taken from reference (Figliozzi et al, 2005).

We show in Fig.7.a the calculated curve for $\alpha = 9°$, when the optical axis is assumed to be perpendicular to sample surface. If compared with the theoretical one, the experimental curve appears to be shifted towards higher ϕ_2. This difference between experimental and theoretical curves suggest that the optical axis may be averagely tilted with respect to the surface normal. From the point of view of the investigated ZnO film, this is a reasonable assumption, taking into account the low temperature deposition technique which was employed. In order to fit the experimental data, an angular tilt of the optical axis was then introduced in the analytical model through a rotation matrix, applied on the $d_{eff}(\alpha)$. This rotation produces the arising of some new terms in the nonlinear optical tensor. In Fig. 7.b we show the polarization chart calculated in this way, assuming a tilt of only 2 around the x-axis, as shown in Fig.8.

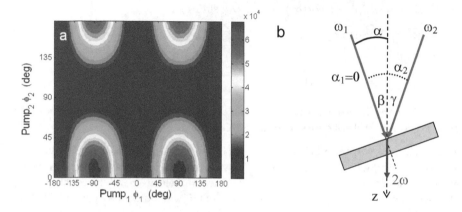

Fig. 6. (a) Noncollinear SH signal measured as a function of the polarization state of the first pump beam (ϕ_1) and the second pump beam (ϕ_2).The polarization state of the analyzer is set to $\hat{s}$, i.e. ϕ=90°. (b) Sketch of the experimental configuration: sample rotation angle was fixed to α=9°.

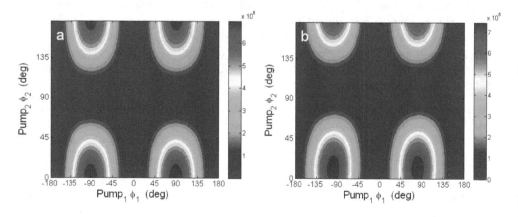

Fig. 7. Theoretically calculated curves of $\hat{s}$ -polarized second harmonic signal as a function of the polarization angle of the first pump beam (ϕ_1) and the second pump beam (ϕ_2), calculated for the optical axis (a) normal to the sample surface and (b) tilted about the x-axis of 2 degrees. Sample rotation angle is α= 9° .

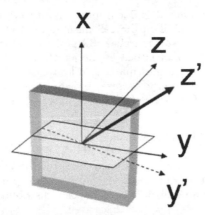

Fig. 8. Sketch of the film orientation. z and z′ represents the optical axis orientation before and after rotation about the x-axis, respectively.

The obtained theoretical curve displays the same ϕ_z-shift evidenced in the experimental curves, thus confirming that the film has a partially oriented polycrystalline structure, as also shown by the X-ray analysis, but the orientation of the optical axis is not exactly normal to the film surface. Similar curves were calculated by tilting the optical axis, along with the nonlinear optical tensor, around the other two reference axes. It's worth to note that for the investigated crystalline symmetry group, 6mm, a rotation about the z-axis does not produce any change in the nonlinear optical tensor. On the other side, a rotation about the y-axis produce an analogous shift in the $\hat{s}$ -polarized SH pattern, but also a modification in the $\hat{p}$ -polarized SH pattern which was not compatible with the corresponding experimental curves.

We conclude, from the experimental results obtained from ZnO films deposited by dual ion beam sputtering, that the *polarization chart* of the noncollinear SH signal can provide important information on the crystalline structure of the films. Specifically, the polarization scanning method adopted is a valid and sensitive tool to probe the orientation of the optical axis and to evidence possible angular tilt with respect to surface normal.

4. Application of the nonlinear ellisometry to Bacteriorhodopsin films

We recently extended the use of the nonlinear ellisometry to the study of chiral molecules, i.e. those molecules lacking an internal plane of symmetry thus having a non-superimposable mirror image. SHG processes have been extensively used for the characterization of optical chirality, due to the large obtainable effects, with respect to conventional linear optical techniques. Considering the nonlinear optical tensor, in fact, the optical chirality is responsible for the introduction of the so-called chiral components. The study of optical chirality by means of SHG was first introduced by Petralli-Mallow and co-workers from a circularly polarized fundamental beam (Petralli-Mallow et al, 1993). Later on, it was demonstrated that also a linearly polarized fundamental beam can be employed to discern chiral components of the nonlinear optical tensor (Verbiest et al, 1995). More recently, a new technique, based on the use of focused laser beams at normal incidence, was applied (Huttunen et al, 2009) to avoid the coupling of possible anisotropy of the sample and thus spurious signals.

The chiral molecule we investigated is Bacteriorhodopsin (BR), a trans-membrane protein found in purple membrane patches in the cell membrane of Halobacterium salinarium, a naturally occurring archaeon in salt marshes. BR proteins naturally arrange in trimers to form a hexagonal two-dimensional lattice in the purple membrane, as shown in Fig.9.a, acting as a natural photonic band gap material (Clays et al, 1993). Furthermore, each BR monomer contains a covalently bound retinal chromophore, presenting its own transition dipole, which is responsible for its outstanding nonlinear optical response (Verbiest et al, 1994) as well as for optical chirality (Volkov et al, 1997).

We examined a 4 μm thick BR film, deposited via an electrophoretic deposition technique onto a substrate covered by a 60 nm thick ITO film. In the resulting BR film, composed by ~800 purple membrane layers (of 5nm thickness each), the chromophore retinal axis is oriented at an angle of 23 ± 4° with respect to the plane of the purple membrane (Schmidt & Rayfield, 1994), i.e. forming an isotropic conical polar distribution around the normal, as shown in Figure 9.b.

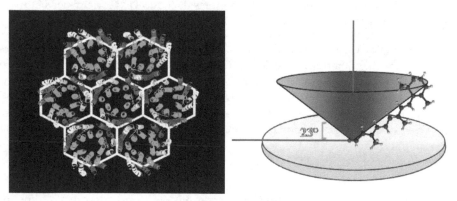

Fig. 9. (a) Hexagonal two-dimensional lattice of BR proteins trimers', as naturally arrange in the purple membrane. (b) Orientation of the retinal chromophores, forming a cone around the normal to the membrane plane, at an angle of 23± 4° relative to the membrane plane.

The BR symmetry structure, arising by consecutive stacking of the naturally hexagonal lattice represented by the membrane sheets having P3 symmetry is noncentrosymmetric, thus its second order susceptibility tensor has three nonvanishing components, i.e. $d_{15}=d_{24}$, $d_{31}=d_{32}$ and d_{33}. Two additional nonzero components of the nonlinear susceptibility tensor, $d_{14}= -d_{25}$, determine the so-called chiral contribution to the nonlinear optical response, since they appear only if molecules have no planes of symmetry (Hecht & Barron, 1996). As a result, the nonlinear optical tensor turns out to be:

$$\tilde{d} = \begin{pmatrix} 0 & 0 & 0 & d_{14} & d_{15} & 0 \\ 0 & 0 & 0 & d_{24} & d_{25} & 0 \\ d_{31} & d_{32} & d_{33} & 0 & 0 & 0 \end{pmatrix} \tag{6}$$

Experimental investigation of the noncollinear SH signal, was performed with the two pump beams angles set to $\beta = 3°$ and $\gamma = -3°$, respectively, while α was fixed to $-40°$. The polarization state of both pump beams was systematically varied in the range $-90° -+90°$. Measurements corresponding to $\hat{p}$ - and $\hat{s}$ -polarized SH are shown in Fig.10.

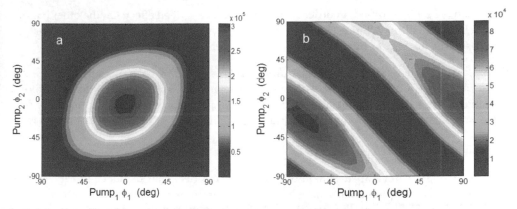

Fig. 10. (a) Noncollinear second harmonic signal experimentally measured as a function of the polarization state of the first pump beam (ϕ_1) and the second pump beam (ϕ_2). Sample rotation angle was fixed to α= -40°. The polarization state of the analyzer is set to (a) $\hat{p}$ i.e. ϕ=0° and (b) $\hat{s}$, i.e. ϕ=90°.

The obtained experimental results indicate that it is possible to retrieve important information also about the optical chirality of the sample from the polarization charts, and in particular from the $\hat{p}$ -polarized signal. In fact, considering the $\hat{p}$ -polarized signal (Figure 10.a) in absence of optical chirality the maximum signal would be located in correspondence of $\phi_1=\phi_2=0°$, i.e. when both pumps are $\hat{p}$ -polarized. In contrast, the pattern shown in Fig.10.a presents a maximum that is somewhat shifted towards the negative quadrant, i.e. both ϕ_1 and ϕ_2 are <0 . It's worth to note that a small value of the chiral components, as can be for instance $|d_{14}|=|d_{25}|= 0.1 \cdot d_{33}$ (Larciprete et al, 2010), still determines an observable effect onto the polarization chart. In Fig.11 the central area of

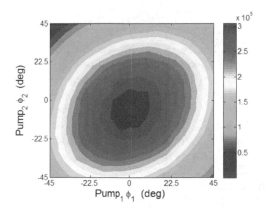

Fig. 11. Detail of the noncollinear second harmonic signal experimentally measured as a function of the polarization state of both pump beams, i.e. ϕ_1 and ϕ_2. Sample rotation angle was α= -40° . The polarization state of the analyzer is set to $\hat{p}$ i.e. ϕ=0°.

Fig.10.a. has been magnified, in order to appreciate the described effect. On the other hand, it must be said that for achiral structures, being $d_{14}=d_{25}=0$ the resulting polarization chart would show a maximum signal well-centred onto the axis' origin.

Finally, from further experimental measurements, performed by changing other parameters as for instance pump beams' power, we trust to put in evidence the role of different potential sources of the nonlinear polarization and in particular the nonlinear magnetic one. This part of the work is still in progress and thus it will be accurately described elsewhere.

5. Conclusion

In conclusion, we developed a method, based on the detection of noncollinear SHG by continuously varying the polarization state of *both* the fundamental beams within a certain range. We have shown that the resulting *polarization charts*, that can be recorded for a given polarization state of the SH signal, present a typical pattern being a signature of a characteristic crystalline structure. First of all, this kind of nonlinear ellipsometry, that doesn't require sample rotation, offers the possibility to evaluate the ratio between the different non-zero elements of the nonlinear optical tensor, or even their absolute values. Furthermore, it represents a valid and sensitive tool to investigate the orientation of the optical axis of a given crystalline structure, being able to evidence possible angular tilt with respect to surface normal. Finally, the polarization scanning method adopted is also able to put in evidence optical chirality, since the so called chiral components of the nonlinear optical susceptibility also introduce some changes in the polarization charts.

6. Acknowledgment

Authors are grateful to Adriana Passaseo (CNR-NNL-INFM Unità di Lecce) and F. Sarto (Division of Advanced Physics Technologies of ENEA, Roma, Italy) for GaN and ZnO sample preparation, respectively.

7. References

Blachnik R.; Chu J.; Galazka R. R.; Geurts J.; Gutowski J.; Hönerlage B.; Hofmann D.; Kossut J.; Lévy R.; Michler P.; Neukirch U.; Story T.; Strauch D.; Waag A. (1999). Zinc oxide (ZnO). *Landolt-Börnstein-Group III Condensed Matter, Semiconductors: II-VI and I-VII Compounds; Semi-magnetic Compounds*, 41B, 52-53, Springer-Verlag. ISBN: 978-3-540-64964-9

Bloembergen N.; Chang R. K.; Jha S. S. & Lee C. H. (1968). Optical Second-Harmonic Generation in Reflection from Media with Inversion Symmetry. *Physical Review*, Vol.174, No.3, pp. 813-822, ISSN: 0031-899X

Bovino F.A.; Larciprete M.C.; Belardini A.; Sibilia C. (2009). Evaluation of the optical axis tilt of zinc oxide films via noncollinear second harmonic generation. *Applied Physics Letters*, Vol.94, No.25, pp.251109 (3 pages), ISSN: 0003-6951

Cattaneo S. & Kauranen M. (2003). Determination of second-order susceptibility components of thin films by two-beam second-harmonic generation. *Optics Letters*, Vol.28, No.16, pp.1445-1447, ISSN: 0146-9592

Cattaneo S. & Kauranen M. (2005). Polarization-based identification of bulk contributions in nonlinear optics. *Physical Review B* Vol.72, No.3, pp.033412 (4 pages) ISSN 1098-0121

Chen J.; Levine Z.H.; Wilkins J.W. (1995). Calculated second harmonic susceptibilities of BN, AlN, and GaN. *Applied Physics Letters*, Vol.66, No.9, pp. 1129-1131, ISSN: 0003-6951

Clays K.; Van Elshocht S. ; Chi M.; Lepoudre E. & Persoons A. (2001). Bacteriorhodopsin: a natural, efficient (nonlinear) photonic crystal. *Journal of the Optical Society of America B*. Vol.18, No.10, pp.1474-1482, ISSN: 0740-3224

Faccio D.; Pruneri V. & Kazansky P.G. (2000). Noncollinear Maker's fringe measurements of second-order nonlinear optical layers. *Optics Letters*, Vol. 25, No.18, pp.1376-78, ISSN: 0146-9592

Figliozzi P.; Sun L.; Jiang Y.; Matlis N.; Mattern B.; Downer M.C.; Withrow S.P.; White C.W.; Mochán W.L. & Mendoza B.S. (2005). Single-Beam and Enhanced Two-Beam Second-Harmonic Generation from Silicon Nanocrystals by Use of Spatially Inhomogeneous Femtosecond Pulses. *Physical Review Letters*, Vol.94, No.4, pp.047401/1-047401/4, ISSN: 0031-9007

Hecht L. & Barron L.D. (1996). New aspects of second-harmonic optical activity from chiral surfaces and interfaces. *Molecular Physics*, Vol.89, No.1, pp.61-80, ISSN: 0026-8976

Huttunen M.J.; Erkintalo M. & Kauranen M. (2009). Absolute nonlinear optical probes of surface chirality. Journal of Optics A: Pure and Applied Optics, Vo.11, No.3, pp.034006, ISSN: 1464-4258

Jerphagnon J. & Kurtz S.K. (1970). Maker fringes: a detailed comparison of theory and experiment for isotropic and uniaxial crystals. *Journal of Applied Physics*, Vol.41, No.4, pp.1667-1681, ISSN: 0021-8979

Larciprete M.C.; Belardini A.; Sibilia C.; Saab M.-b.; Váró G. & Gergely C. (2010). Optical chirality of bacteriorhodopsin films via second harmonic Maker's fringes measurements. *Applied Physics Letters*, Vol.96, No.22, pp.221108 (3 pages), ISSN: 0003-6951

Larciprete M.C.; Bovino F.A.; Giardina M.; Belardini A.; Centini M.; Sibilia C.; Bertolotti M.; Passaseo A. & Tasco V. (2009). Mapping the nonlinear optical susceptibility by noncollinear second-harmonic generation. *Optics Letters*, Vol.34, No.14, pp.2189-2191, ISSN: 0146-9592

Maker P. D.; Terhune R. W.; Nisenoff M. & Savage C. M. (1962). Effects of Dispersion and Focusing on the Production of Optical Harmonics. *Physical Review Letters*, Vol.8, No.1, pp.21-22, ISSN: 0031-9007

Miragliotta J.; Wickenden D.K.; Kistenmacher T.J. & Bryden W.A (1993). *Journal of the Optical Society of America B*, Vol.10, No.8, pp.1447-1456, ISSN: 0740-3224

Muenchausen R.E.; Keller R.A. & Nogar N.S. (1987). Surface second-harmonic and sum-frequency generation using a noncollinear excitation geometry. *Journal of the Optical Society of America B,*, Vol.4, No.2, pp.237-241, ISSN: 0740-3224

Petralli-Mallow T.; Wong T.M.; Byers J.D.; Lee H.I. & Hicks J.M. (1993). Circular dichroism spectroscopy at interfaces: a surface second harmonic generation study. *Journal of Physical Chemistry*, Vol.97, No.7, pp.1383-1388, ISSN: 0022-3654

Poti B.; Tagliente M.A. & Passaseo A. (2006). *Journal of Non-Crystalline Solids*, Vol.352, No.23, pp.2332-2334, ISSN: 0022-3093

Provencher P.; Côté C.Y. & Denariez-Roberge M.M. (1993). Surface second-harmonic susceptibility determined by noncollinear reflected second-harmonic generation. *Canadian Journal of Physics.*, Vol.71, No.1-2, pp.66-69, ISSN: 0008-4204

Schmidt P. K. & Rayfield G. W. (1994). Hyper-Rayleigh light scattering from an aqueous suspension of purple membrane. *Applied Optics*, Vol.33, No.19, pp.4286-4292, ISSN: 1559-128X

Verbiest T.; Kauranen M.; Maki J.J.; Teerenstra M.N.; Schouten A.J.; Nolte R.J.M. & Persoons A. (1995). Linearly polarized probes of surface chirality. Journal of Chemical Physics, Vol.103, No.18, pp.8296- 8298, ISSN. 0021-9606

Verbiest T.; Kauranen M.; Persoons A.; Ikonen M.; Kurkela J. & Lemmetyinen H. (1994). Nonlinear Optical Activity and Biomolecular Chirality. *Journal of the American Chemical Society*, Vol.116, No.20, pp.9203-9205, ISSN: 0002-7863

Volkov V.; Svirko Yu. P.; Kamalov V. F.; Song L. & El-Sayed M. A. (1997). Optical rotation of the second harmonic radiation from retinal in bacteriorhodopsin monomers in Langmuir-Blodgett film: evidence for nonplanar retinal structure. *Biophysical Journal*, Vol.73, No.6, pp.3164-3170, ISSN: 0006-3495

Weißenrieder K.S. & Muller J. (1997). Conductivity model for sputtered ZnO-thin film gas sensors. *Thin Solid Films*, Vol.300, No.1-2, pp.30-41, ISSN: 0040-6090

5

Donor-Acceptor Conjugated Polymers and Their Nanocomposites for Photonic Applications

D. Udaya Kumar[1], A. John Kiran[2], M. G. Murali[1] and A. V. Adhikari[1]
[1]Department of Chemistry, National Institute of Technology Karnataka,
Surathkal, P. O. Srinivasnagar,
[2]Department of Inorganic and Physical Chemistry,
Indian Institute of Science, Bangalore
India

1. Introduction

Organic materials exhibiting strong nonlinear optical (NLO) properties have attracted considerable interest in recent years because of their promising applications in opto-electronic and all-optical devices such as optical limiters, optical switches and optical modulators (Munn & Ironside, 1993; Zyss, 1994). A variety of organic materials, including conjugated molecules, polymers and dyes, have been investigated for their NLO responses (Kamanina, 1999, 2001; Kamania & Plekhanov, 2002; Kamanina et al., 2008, 2009). Conjugated organic polymers have emerged as a promising class of NLO materials because of their large nonlinear responses associated with fast response time, in addition to their structural variety, processability, high mechanical strength, and excellent environmental and thermal stability (Prasad & Williams, 1992). In contrast to misconceptions about the frailty of simple organic molecules, the optical damage threshold for polymeric materials can be greater than 10 GW/cm². Among various π-conjugated materials, thiophene based polymers are currently under intensive investigation as materials for nonlinear optics because of their large third-order nonlinear response, chemical stability and their readiness of functionalization (Kishino et al., 1998; Nisoli et al, 1993; Sutherland, 1996, Udayakumar et al., 2006).

It has been well-known that, the strong delocalization of π-electrons along the backbone of conjugated polymers determines very high molecular polarizability and thus causes remarkable optical nonlinearities. However, a necessary step in further improving the NLO properties of conjugated polymers is to understand the fundamental relationship that exists between the molecular structure and the hyperpolarizabilities. A deeper understanding in this subject will improve the design of organic conjugated molecules and polymers by a judicious choice of functional substituents to tune their optical properties for photonic applications. Cassano et al. had reported a strategy for tuning the linear and nonlinear optical properties of soluble poly(paraphenylenevinylene) derivatives, based on the effect of simultaneous presence of electron-acceptor and electron-donor units in the conjugated backbone (Cassano, 2002). Particularly, they reported that the value of third-order nonlinear susceptibility ($\chi^{(3)}$) obtained for the polymer containing alternating phenylenevinylene and tetrafluorophenylenevinylene units was one order of magnitude larger than that measured for the corresponding homopolymer, poly(paraphenylenevinylene) (PPV). Similarly, Chen et al. reported the third-order optical nonlinearity of a conjugated 3,3' - bipyridine derivative, with an enhanced nonlinearity due to its symmetrical donor-acceptor-donor structure, in which the 3,3' - bipyridine core was acting as the acceptor group and the thiophene ring as the donor group (Chen, 2003). The donor-acceptor (D-A) approach, introduced by Havinga et al. (1993) in macromolecular systems via alternating electron-rich and electron-deficient substituents along the conjugated backbone, has attracted a good deal of attention in recent years. Interaction of the donor–acceptor moieties enhances the double bond character between the repeating units, which stabilizes the low band gap quinonoid like forms within the polymer backbones. Hence, a conjugated polymer with an alternating sequence of the appropriate donor and acceptor units in the main chain can induce a reduction in its band gap energy. Recently, molecular orbital calculations have shown that the hybridization of the energy levels of the donor and the acceptor moieties result in D-A systems with unusually low HOMO-LUMO separation (Brocks & Tol, 1996). If the HOMO levels of the donor and the LUMO levels of the acceptor moiety are close in energy, the resulting band structure will show a low energy gap. Further reduction in band gap is possible by enhancing the strength of donor and acceptor moieties via strong orbital interactions. In addition, the presence of strong electron donors and acceptors in conjugated polymers increases the π-electron delocalization leading to a high molecular polarizability and thus improving the nonlinear response of the polymer. In this direction, several donor-acceptor type conjugated polymers were synthesized and their third-order nonlinear optical properties were investigated (Hegde et al., 2009; Kiran et al., 2006, Udayakumar et al., 2006, 2007).

Metal and semiconductor nanoparticles are also emerging as a promising class of NLO materials for nanophotonic applications (Fatti & Vallee, 2001; Gayvoronsky et al., 2005; Philip et. al., 2000; Venkatram et. al., 2005; Voisin et al., 2001). For instance, a large nonlinear optical response with a third-order NLO susceptibility ($\chi^{(3)}$) value as high as 2×10^{-5} esu has been observed for nanoporous layers of TiO_2 (Gayvoronsky et al., 2005). Research has shown that it is advantageous to embed metal/semiconductor nanoparticles in thin polymer films for application purposes because the polymer matrix serves as a medium to assemble the nanoparticles and stabilize them against aggregation (Boyd, 1996; Takele et al., 2006). Furthermore, the nanocomposite structures are known to substantially enhance the optical nonlinearities (Neeves & Birnboim 1988). Therefore, the third-order NLO properties of several metal/semiconductor-polymer nanocomposites have been investigated (Gao et al., 2008; Karthikeyan et al., 2006; Porel et al., 2007). Recently, conjugated polymers are being used as host matrix for dispersing metal/semiconductor nanoparticles (Sezer et al., 2009). Such nanoparticle/polymer composites are shown to possess a large third-order nonlinear susceptibility of the order of 10^{-7} esu with an ultrafast response time of 1.2 ps (Hu et al., 2008). The nanocomposites made of Ag nanoparticles dispersed in poly[2-methoxy-5-(2-ethylhexyloxy)-1,4-phenylenevinylene] matrix have exhibited a large third-order nonlinear susceptibility of the order of 10^{-6} esu (Hu et al., 2009). Higher $\chi^{(3)}$ value has been observed for polydiacetylene-Ag nanocomposite film when compared with the pure polydiacetylene film (Chen et al., 2010).

In this chapter, we describe the third-order nonlinear optical studies on new donor-acceptor conjugated polymers; named as PThOxad 1a-c, PThOxad 2a-c and PThOxad 3a-c. All these polymers contain 3,4-diakoxythiophene units as electron donors and 1,3,4-oxadiazole as electron acceptor units. Nanocmposites of PThOxad 3c and nano TiO_2 were prepared using the spin coating method. The third-order nonlinear optical properties of the polymers and their composites were investigated using z-scan and degenerate four wave mixing (DFWM) techniques. All the polymers exhibit significant third-order nonlinearity. We describe their concentration dependent optical limiting behavior using a ns Nd:YAG laser operated at 532 nm.

2. Experimental set up

Z-scan (Bahae et al., 1990) is a technique that is particularly useful when nonlinear refraction is accompanied by nonlinear absorption. This method allows the simultaneous measurement of both the nonlinear refractive index and the nonlinear absorption coefficient of a material. Basically, the method consists of translating a sample through the focus of a Gaussian beam and monitoring the changes in the far-field intensity pattern. Because of the light-induced lens-like effect, the sample has a tendency to recollimate or defocus the incident beam, depending on its z position with respect to the focal plane. By properly monitoring the transmittance change through a small aperture placed at the far-field position (closed aperture), one is able to determine the amplitude of the phase shift. By moving the sample through the focus and without placing an aperture at the detector (open aperture), one can measure the intensity-dependent absorption as a change of transmittance through the sample.

A Q-switched Nd:YAG laser with a pulse width of 7 ns at 532 nm was used as a source of light in the z-scan experiment. The output of the laser had a nearly Gaussian intensity profile. A lens of focal length 26 cm was used to focus the laser pulses onto the sample. The resulting beam waist radius at the focused spot, calculated using the formula $w_0 = 1.22\lambda f / d$, where f is focal length of the lens and d is the diameter of the aperture, was found to be 20 μm. The corresponding Rayleigh length, calculated using the formula, $z_0 = \pi w_0^2/\lambda$ was found to be 2.3 mm. Thus the sample thickness of 1 mm was less than the Rayleigh length and hence it could be treated as a thin medium. The scan was obtained with a 50% (S = 0.5) aperture and at pulse energy of 10 μJ, which corresponds to a peak irradiance of 0.22 GW/cm^2. In order to avoid cumulative thermal effects, data were collected in single shot mode (Yang, 2002). The optical limiting measurements were carried out when the sample was at focal point by varying the input energy and recording the output energy. Both the incident and the transmitted energies were measured simultaneously by two pyroelectric detectors with Laser Probe Rj-7620 Energy Ratio meter. Spectral grade dimethylformamide (DMF) was used for the preparation of polymer solutions.

Four-wave mixing refers to the interaction of four waves in a nonlinear medium via the third-order polarization. When all the waves have same frequency, it is called as degenerate four-wave mixing. There are several geometries used in studying this phenomenon. One of such geometries used in our experiment is the backward geometry or the phase conjugate geometry. Here, two counter propagating strong beams are called forward pump beam and the backward pump beam. A third wave called the probe beam is incident at small angle θ (~ 4°) to the direction of the forward pump. A fourth beam, called the conjugate beam, is generated in the process and propagates counter to the probe beam (Sutherland, 1996). In the present study, the laser energy (at 532 nm) at the sample was varied by the combinations of neutral density filters. Sample was taken in a 1 mm thick glass cuvette, with a concentration of 10^{-5} mol/L. A small portion of the pump beams was picked off and measured by a photodiode to monitor the input energy. The DFWM signal generated in the sample solution was separated by a second photodiode. The photodiode signals were averaged over a number of laser shots and displayed by a Tektronix TDS2002 digital storage oscilloscope.

3. Polymers and their nanocomposites

Introduction of strong electron donor and strong electron acceptor groups along polymer chain could be a promising molecular design to improve the NLO properties in D-A type polymers. Processing of these polymers for application purposes requires good solubility in common organic solvents. Another important criterion from the application point of view is the good film forming properties of these polymers. Solubility of the polymers could be improved by the incorporation of proper solubilising groups either in the polymer main chain or in the side chain. Keeping these points in view, three series of D-A polymers (PThOxad 1a-c, PThOxad 2a-c and PThOxad 3a-c) are synthesized in the present study and are characterized. The polymer structures consist of electron donating 3,4-dialkoxythiophene units and electron accepting 1,3,4-oxadiazole units. In 3,4-dialkoxythiophnes, introduction of long alkoxy pendants at 3- and 4- positions of the thiophene ring not only enhances the electron donating nature of the ring but also improves the solvent processability of the corresponding polymer. On the other hand, 1,3,4-oxadiazole ring, due to its high electron affinity, is a good electron acceptor for D-A type conjugated polymers. However, in the case of poly(3,4-dialkoxythiophene)s the steric interactions of alkoxy groups of adjacent thiophene rings reduce the coplanarity and hence it affects the conjugation length of the polymer. In order to minimize such steric interactions two strategies have been employed in the present study. First one is to introduce a spacer unit, like a 1,4-divinylbenzene moiety as is the case in PThOxad 1a-c (Fig. 1) or a phenyl ring (PThOxad 3a-c), along the polymeric backbone so that the thiophene rings are well separated, which thus minimizes the steric interactions of the alkoxy groups. Second method is to replace one of the 3,4-dialkoxythiophene units by a cyclosubstituent fused at 3- and 4-positions, i.e. 3,4-ethylenedioxythiophene (EDOT) unit (PThOxad 2a-c), so that there will not be any steric interactions. These polymers are synthesized by using a precursor polyhydrazide route. As expected, all these polymers showed good solubility in common organic solvents, which is an important requirement for the processing of the polymers for device applications.

The chemical structures of the D-A polymers containing 3,4-dialkoxythiophenes and 1,3,4-oxadiazole units are shown in Fig. 2. The synthetic route for these polymers involves the preparation of monomer units, which follows the polycondensation of these monomeric units to give the final polymers. As a representative, the synthesis of PThOxad 1a-c is described in this section. A series of monomers, 3,4-dialkoxythiophene-2,5-carbonyldihydrazides (6a-c) were prepared starting from thiodiglycolic acid (1). Esterification reaction of thiodiglycolic acid (1) with ethanol in presence of conc.sulphuric acid afforded diethylthiodiglycolate (2). Compound 2 was then condensed with diethyloxalate in presence of sodium ethoxide and ethanol to get diethyl 3,4-dihydroxythiophene-2,5-dicarboxylate disodium salt (3). Acidification of the disodium salt with hydrochloric acid afforded the compound 4.

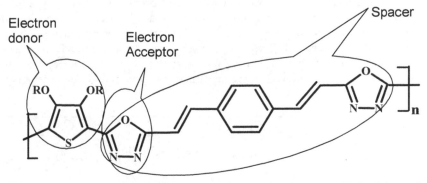

Fig. 1. Chemical structures of donor-acceptor conjugated polymers, PThOxd 1a-c, showing donor, acceptor and spacer units.

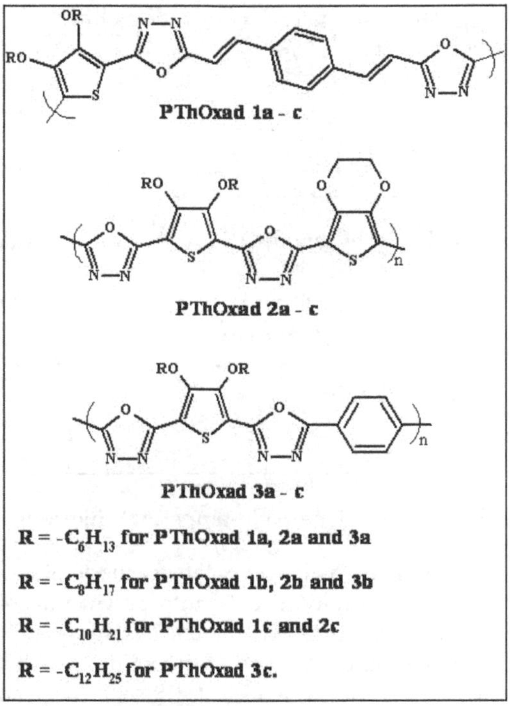

PThOxad 1a - c

PThOxad 2a - c

PThOxad 3a - c

R = -C_6H_{13} for PThOxad 1a, 2a and 3a

R = -C_8H_{17} for PThOxad 1b, 2b and 3b

R = -$C_{10}H_{21}$ for PThOxad 1c and 2c

R = -$C_{12}H_{25}$ for PThOxad 3c.

Fig. 2. General structures of the polymers, PThOxad 1a-c, PThOxad 2a-c and PThOxad 3a-c.

Diethyl 3,4-dialkoxythiophene-2,5-dicarboxylates (5a-c) were synthesized by the etherification reaction of diethyl 3,4-dihydroxythiophene-2,5-dicarboxylate (4) with the corresponding n-bromoalkane in the presence of potassium carbonate and DMF. The reaction was completed in 70 h. Diethyl 3,4-dialkoxythiophene-2,5-dicarboxylates (5a-c) were then converted into corresponding 3,4-dialkoxythiophene-2,5-carbonyldihydrazides (6a-c) by treating them with hydrazine hydrate and methanol. Scheme 2 shows synthetic route to the synthesis of 3,3'-(1,4-phenylene)bis[2-propenoyl chloride] (8), which was achieved starting from 1,4-benzenedicarboxaldehyde. Knoevenagel condensation of 1,4-benzenedicarboxaldehyde (7) with malonic acid in presence of pyridine and a catalytic amount of piperidine afforded 3,3'-(1,4-phenylene)bis[2-propenoic acid], which on treatment with thionyl chloride utilizing DMF as a catalyst yielded the monomer 8. The chemical structures of all the above compounds were confirmed by spectral and elemental analyses.

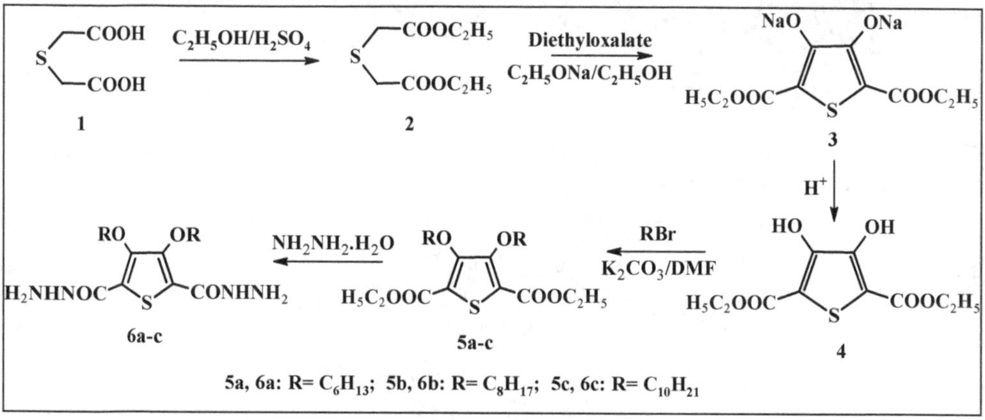

Scheme 1. Synthesis of 3,4-dialkoxythiophene-2,5-carbonyldihydrazides (6a-c).

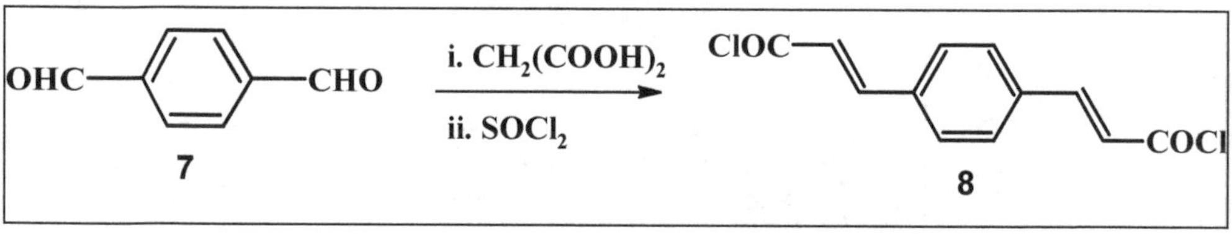

Scheme 2. Synthesis of 3,3'-(1,4-phenylene)bis[2-propenoyl chloride] (8)

The synthetic route for PThOxad 1a-c is involves the synthesis of precursor polyhydrazides followed by the conversion these polyhydrazides into polyoxadiazoles (scheme 3). For the preparation of polyhydrazides, a mixture of 10 mmol of appropriate dihydrazide, 20 mmol of anhydrous lithium chloride and 0.1 ml of pyridine was taken in 20 ml of N-methylpyrrolidinone, and 10 mmol of acid chloride (8) was added slowly at room temperature under N_2 atmosphere. The reaction mixture was stirred at room temperature for 5 h. The resultant yellow solution was heated at 80 °C with stirring for 20 h. After cooling to room temperature the reaction mixture was poured into water to get a precipitate. The precipitate was collected by filtration and was washed thoroughly with water followed by acetone and finally dried in vacuum to get the corresponding polyhydrazides in 70-85% yield. These polyhydrazides are insoluble in common organic solvents at ambient and even at elevated temperatures, as reported for other polyhydrazides. The polyhydrazides were converted into the corresponding poly(1,3,4-oxadiazole)s, by cyclodehydration of the hydrazide group into 1,3,4-oxadiazole ring, using polyphosphoric acid (PPA), which functions both as solvent and dehydrating agent. The reaction was carried out by heating a mixture of polyhydrazide (0.5 g) and 50 ml of polyphosphoric acid at 100 °C for 4 h under N$_2$ atmosphere. The reaction mixture was then cooled to room temperature and poured into excess of water. The resulting precipitate was collected by filtration, washed thoroughly with water followed by acetone and dried in oven at 70 °C to get the final polymers in 70 - 80% yield. The progress of the cyclodehydration reaction was monitored by FTIR spectroscopy. The stretching bands of C = O and N – H groups of polyhydrazides (fig.3), disappeared in the FTIR spectra of the corresponding poly(oxadiazole)s, where as the band corresponding to imine (C = N) in an oxadiazole ring was newly generated (fig. 4). In addition, peaks due to =C–O–C= (1,3,4-oxadiazole ring) stretching was also observed for these polymers, confirming the conversion of polyhydrazides to polyoxadiazoles.

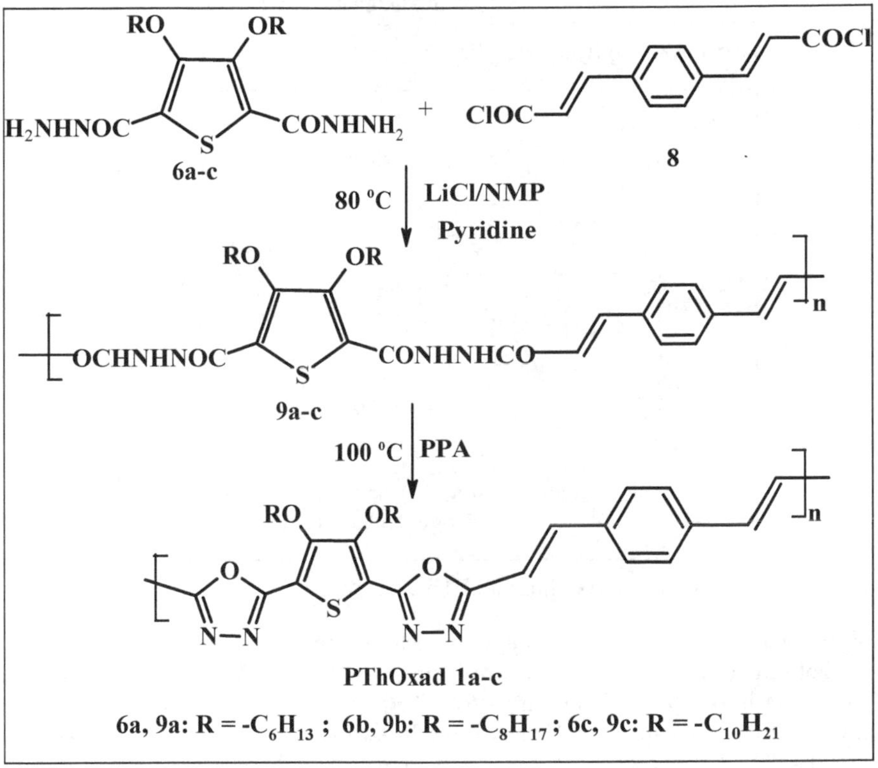

Scheme 3. Synthesis of polymers, PThOxad 1a-c.

The polymers are found to be thermally stable up to ~330 °C. DSC studies were performed to observe glass transition temperature (T$_g$) of the polymers. The samples were heated up to 300 °C under nitrogen atmosphere at a heating rate of 5 °C/min. No T$_g$ or melting point was observed suggesting that the polymers are either having very high T$_g$ or are highly crystalline in nature and decompose before melting.

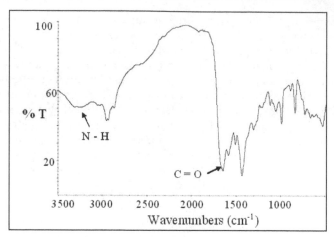

Fig. 3. FT-IR sprectrum of polyhydrazide 9c.

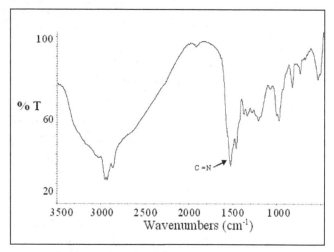

Fig. 4. FT-IR sprectrum of polyoxadiazole, PThOxad 1c.

The UV-vis absorption and fluorescence spectra of the polymers (PThOxad 1a-c) were recorded both in solution and in thin film form. As shown in fig. 5, the absorption maxima of the polymers in dilute DMF solutions (ca. 10^{-5}) are 373 nm for PThOxad 1a, 378 nm for PThOxad 1b and 381 nm for PThOxad 1c. In addition, the absorption spectra of the polymers displayed a shoulder at 306 nm. Compared with poly(3,4-dialkoxythiophenes), these polymers showed a red shift in absorption spectra. This may be due to the presence of 1,4- divinylbenzene moiety in PThOxad 1a-c, which serves to alleviate steric effects of the alkoxy groups in the adjacent thienylene rings. Hence, electron-donating contributions from the alkoxy groups to the electronic structure of the polymers become more prominent. The absorption spectra of the polymer thin films (fig. 6) are rather broad and so their λ_{max} values could not be precisely determined. However, their optical energy band gap (E_g) was calculated from the absorption edge in the thin films to be 2.20–2.38 eV. As shown in fig. 7,

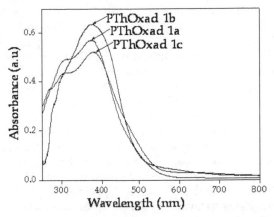

Fig. 5. UV-visible absorption spectra of PThOxad 1a-c in DMF solution.

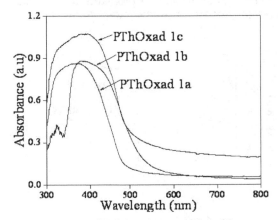

Fig. 6. UV-visible absorption spectra of PThOxad 1a-c as thin films.

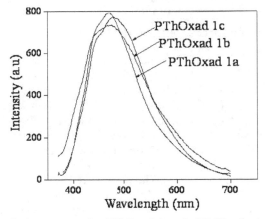

Fig. 7. Fluorescence emission spectra for PThOxad 1a-c in DMF solution.

the emissive maxima of the polymers in dilute DMF solutions (ca. 10^{-5}) are 470 nm for PThOxad 1a, 474 nm for PThOxad 1b, and 479 nm for PThOxad 1c. The Stokes shift was determined to be 97, 96 and 98 nm for PThOxad 1a-c, respectively. The fluorescence emission spectra of these polymers in thin films are shown in fig. 8. The polymers emit intense green light in solid state, with emission peaks at 502, 506 and 512 nm for PThOxad 1a-c, respectively. Consequently, the fluorescence spectra of the polymer thin films exhibit a red shift with respect to those obtained from their solutions. This can be attributed to the interchain or/and intrachain mobility of the excitons and excimers generated in the polymer in the solid stated phase. Further, a sequential red shift in the λ_{max} was observed in both the UV–vis absorption spectra and fluorescence emission spectra of polymers. The increase in the length of the alkoxy side chains led to a red shift in the λ_{max} in UV–vis absorption and fluorescence emission spectra. This can be interpreted as an expected better side chain interdigitation and interchain organization with increasing pendant chain length. The fluorescence quantum yields (Davey et al., 1995) of the polymers in solution were determined using quinine sulfate as a standard (Demas & Grosby, 1971) and are found to be in the range of 26–30%. Following the similar procedure for PThOxad 1a-c, other two series of polymers, PThOxad 2a-c and PThOxad 3a-c, were prepared and their optical properties were studied.

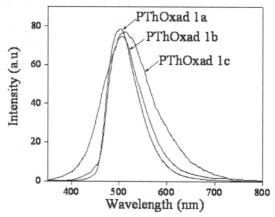

Fig. 8. Fluorescence emission spectra for PThOxad 1a-c as thin films.

For the preparation of PThOxad 3c/TiO$_2$ nanocomposite, polymer (PThOxad 3c) and TiO$_2$ were dispersed in the weight ratio 4:1 in chloroform and chlorobenzene solvent system (10:1 volume ratio) and sonicated for 2 hrs. Nanocomposite films were prepared on clean glass plates through spin coating and the films were dried in vacuum for 1 hr. The thickness of the polymer and the nanocomposite films was determined by the SEM cross section, and was found to be in the range of 0.5 – 1 micrometer. The linear transmittance of the film samples was between 50 to 60 % aqt 532 nm. The thermogravimetric analysis (TGA) of PThOxad 3c and PThOxad 3c/TiO$_2$ nanocomposite was carried out under nitrogen atmosphere at a heating rate of 5 $^\circ$C/min. The polymer (PThOxad 3c) decomposed slowly in the temperature region of 190 – 310 $^\circ$C and thereafter a rapid degradation took place up to 495 $^\circ$C. The nanocomposite started to decompose at 300 $^\circ$C (fig. 9) which indicated a higher thermal stability of the nanocomposite in comparison with that of the polymer. A similar thermal stabilization of nanocomposite was reported in the literature (Zhu et al., 2008). Fig.10 shows the SEM image of the PThOxad 3c/TiO₂ nanocomposite. A moderately uniform distribution of TiO₂ nanoparticles can be observed with average particle sizes ranging from 25 to 45 nm.

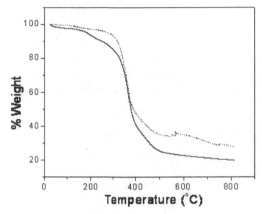

Fig. 9. TGA graphs for PThOxad 3c (—) and PThOxad 3c/TiO₂ nanocomposite (---).

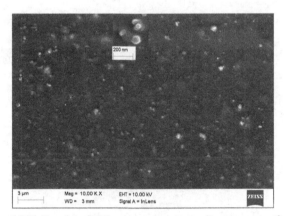

Fig. 10. SEM images of PThOxad 3c/TiO₂ nanocomposite (Inset: magnified image, Mag. = 100 K X).

The UV-vis absorption spectra of PThOxad 3c in CHCl₃ solution (ca.10⁻⁵), PThOxad 3c film and PThOxad 3c/TiO₂ nanocomposite film are shown in fig. 11. PThOxad 3c in solution displayed an absorption maximum at 378 nm, while the PThOxad 3c film showed a red shift of 16 nm in the absorption spectrum. The optical energy band gap (E_g) of PThOxad 3c was calculated from the absorption edge in the film and was found to be 2.28 eV. The PThOxad 3c/TiO₂ nanocomposite film showed an absorption peak at 412 nm (π→ π* of polymer) and at 310 nm and a shoulder at 250 nm (characteristic absorptions of TiO₂). Fig. 12 shows the fluorescence emission spectra for PThOxad 3c in CHCl₃ solution (ca.10⁻⁴ g/l), PThOxad 3c film and PThOxad 3c/TiO₂ nanocomposite film. Incorporation of TiO₂ nanoparticles caused a slight red shift in the absorption spectra and a blue shift in the emission spectra. There was a shift of 18 nm in the absorption maximum of the nanocomposite film, when compared with that of the polymer film. We note that, a similar red shift in the absorption maximum was reported for MEH-PPV/TiO$_2$ nanocomposite films (Yang et al., 2007). A blue shift of 26 nm was observed in the emission maximum of PThOxad 3c/TiO$_2$ nanocomposite film in comparison with that of the polymer film. Although the exact reasons for these shifts are not well understood, these optical results indicate that some interactions can occur between the conjugated polymer chains and TiO$_2$ nanoparticles. It can be suggested that these interactions will modify the polaronic states of the polymer and increase the excitonic energy. Therefore, the emission spectrum of PThOxad 3c/TiO$_2$ nanocomposite will shift toward higher photo energies than those of the PThOxad 3c resulting in a blue shift in the emission spectrum (Hsieh et al., 2007).

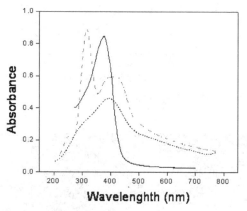

Fig. 11. UV-vis absorption spectra for polymer solution in chloroform (−), polymer film (---) and nanocomposite film (...).

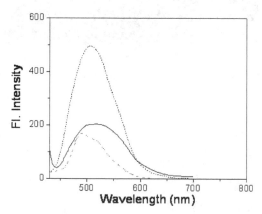

Fig. 12. Fluorescence emission specta of the polymer solution in chloroform (---), polymer film (−) and nanocomposite film (...).

4. Results and discussion

4.1 Polythiophenes

Fig. 13 shows the open aperture z-scan results obtained at 532 nm for PThOxad 1a-c samples dissolved in DMF. The transmission is symmetric about the focus (z = 0) where it has a minimum transmission. Thus an intensity dependent absorption effect is observed. The solid line in Fig. 13 is a fit of data to equation (1), by assuming only two-photon absorption (2PA). The normalized transmission for the open aperture condition (Henari et al., 1997) is given by,

$$T(z) = 1 - \frac{q_0}{2\sqrt{2}} \quad \text{for } q_0 < 1, \tag{1}$$

where:

$$q_0(z) = \frac{\beta \, I \, (1 - \exp^{-\alpha L})}{(1 + z^2/z_0^2)\, \alpha}.$$

Here, β is the nonlinear absorption coefficient, L is the length of the sample, I is the intensity of the laser beam at the focus, z is the position of the sample and z_0 is the Rayleigh range of the lens. A fit of open aperture data with equation (1) yields the value of β in the range of 50 to 80 cm/GW for the polymer samples. We note that, the linear absorption spectra of PThOxad 1a-c (fig. 5) show that the polymers are not fully transparent at 532 nm; the linear absorption coefficients (α) for polymers at 532 nm are tabulated in Table 1. Therefore, there can be various mechanisms, such as the excited state absorption (ESA), that are responsible for such a large nonlinear absorption in these polymers.

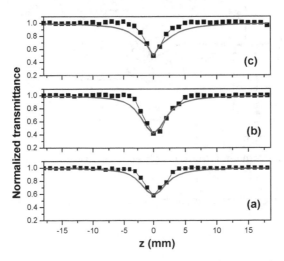

Fig. 13. Open aperture curves for PThOxad 1a (a), PThOxad 1b (b) and PThOxad 1c (c). Solid line is a fit of data to equation (1) assuming only 2PA, with β =73.8 cm/GW for PThOxad 1a, β = 53 cm/GW for PThOxad 1b and β =63.6 cm/GW for PThOxad 1c.

Fig. 14 represents different nonlinear absorption (NLA) mechanisms using an energy level diagram. The diagram shows the different energy levels of a molecule, the singlet ground state S_0, the excited singlet states S_1 and S_2, as well as the triplet excited states T_1 and T_2. It also displays the different transitions taking place between these energy levels. When two photons, of the same or different energy are simultaneously absorbed from the ground state to a higher excited state ($S_0 \rightarrow S_1$), it is denoted as two-photon absorption (2PA). When the excited state absorption (ESA) occurs, molecules are excited from an already excited state to a higher excited state (e.g. $S_1 \rightarrow S_2$ and/or $T_1 \rightarrow T_2$). For this to happen the population of the excited states (S_1 and/or T_1) needs to be high so that the probability of photon absorption from that state is high. The ESA could be enhanced if the molecules could undergo intersystem crossing (ISC) to the triplet state. If more absorption occurs from the excited state than from the ground state it is usually called as the reversed saturable absorption (RSA). The triplet excited state absorption may result in RSA if the absorption cross section of triplet excited state is greater than that of singlet excited state. With excitation of laser pulses on the nanosecond scale, triplet-triplet transitions may make a significant contribution. Nevertheless, such an increased nonlinear absorption due to additional excited state absorption will result in a strong optical limiting activity of the material.

The excited state cross section (σ_{ex}) can be measured from the normalized open aperture z-scan data (Henari et al., 1997). It is assumed that the molecular energy levels can be reduced to a three level system in order to calculate σ_{ex}. Molecules are optically excited from the ground state to the singlet-excited state and from this state they relax either to the ground state or the triplet state, when excited state absorption can occur from the triplet to the higher triplet excited state.

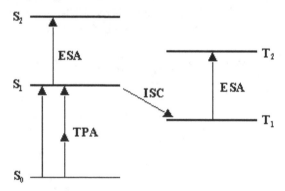

Fig. 14. Energy level diagram showing both two-photon absorption and excited state absorption.

The change in the intensity of the beam as it passes through the material is given by,

$$\frac{dI}{dz} = -\alpha I - \sigma_{ex} N(t) I \,,$$

where I is the intensity and N is the number of molecules in the excited state. The excited state density of molecules appears as a result of a nonlinear absorption process, whose intensity dependence can be obtained from,

$$\frac{dn}{dz} = \frac{\sigma_{ex} I}{h\nu},$$

where ν is the frequency of the laser. Combining the above two equations and solving for the fluence of the laser and integrating over the spatial extent of the beam gives the normalized transmission for open aperture as

$$T = \ln\left(1 + \frac{q_0}{1+x^2}\right) / \frac{q_0}{1+x^2} \tag{2}$$

where $x = z/z_0$ and $q_0 = \dfrac{\sigma_{ex}\alpha F_0 L_{eff}}{2h\gamma}$. Here F_0 is the fluence of the laser at the focus and

$L_{eff} = \dfrac{(1\text{-}exp^{-\alpha L})}{\alpha}$.

The values of σ_{ex} of PThOxad 1a-c obtained through a fit of equation (2) to the corresponding open aperture data at 532 nm with q_0, are tabulated in Table 1. The value of ground state absorption cross-section of the polymers, as calculated from $\alpha = \sigma_g N_a C$, where N_a is Avogadro's number and C is the concentration in moles/cm^3, is also given in Table 1. It is clear that the value of σ_{ex} is larger than the value of σ_g in all the polymers, which is in agreement with the condition for observing the reverse saturable absorption (Henari et al., 1997; Tutt & Boggess, 1993). Reverse saturable absorption generally arises in a molecular system when the excited state absorption cross section is larger than the ground state cross section. The background linear absorption at 532 nm and the measured σ_{ex} values indicate that there is a contribution from excited state absorption to the observed NLA. This suggests that, the nonlinear absorption observed in the polymers can be attributed to a reverse saturable absorption.

Polymer	α (cm^{-1})	σ_g ($\times 10^{-18}$ cm^2)	σ_{ex} ($\times 10^{-17}$ cm^2)
PThOxad 1a	0.0487	8.092	9.66
PThOxad 1b	0.0622	10.32	5.44
PThOxad 1c	0.068	11.26	5.93

Table 1. The values of α, σ_g and σ_{ex} for the copolymers.

The reverse saturable absorption in these polymers can further be verified by plotting the nonlinear absorption coefficient against the incident intensity. Fig. 15 shows the plot of β versus the input intensity for PThOxad 1a-c in DMF at a concentration of 1x10^{-5} mol/L. Generally, nonlinear absorption (NLA) can be caused by free carrier absorption, saturated absorption, direct multiphoton absorption or excited state absorption. If the mechanism belongs to the simple two-photon absorption, β should be a constant that is independent of the on-axis irradiance I_0. If the mechanism is direct three-photon absorption β should be a linear increasing function of I_0 and the intercepts on the vertical axis should be nonzero (Guo et al., 2003). As shown in fig. 15, β is found to decrease with increasing input intensity. Such fall-off of β with increasing intensity is a consequence of the reverse saturable absorption (Couris et al., 1995). With increasing intensity the total absorption of these polymers approaches asymptotically the absorbance of the triplet state. Therefore, the β will be reduced at least up to intensities where no other intensity dependence processes are involved which can further cause reduction of transmission of polymer solution.

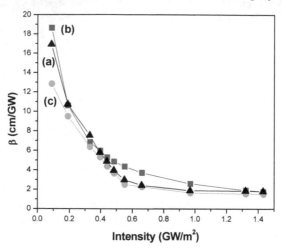

Fig. 15. β versus intensity for PThOxad 1a-c in DMF solution (1×10^{-5}mol/L). The magnitude of β decreased as intensity increased.

Fig. 16 shows the normalized transmission for the closed aperture Z-scan obtained for PThOxad 1a-c. These pure nonlinear refraction curves were obtained through dividing the closed aperture data by the corresponding open aperture data. The z-scan signature shows a large negative refractive nonlinearity (self defocusing) for all the polymers. The closed aperture data was fitted with equation (3) given below (Bahae et al., 1990; Liu et al., 2001):

$$T(z) = 1 - \frac{4\Delta\phi_0 x}{(1+x^2)(9+x^2)} , \qquad (3)$$

where $\Delta\phi_0$ is the phase change given by $\Delta\phi_0 = \dfrac{\Delta T_{p-v}}{0.406(1-S)^{0.25}}$ for $|\Delta\phi_0| \leq \pi$.

The values of nonlinear refractive index (n_2) for PThOxad1a-c were found to be of the order of 10^{-10} esu, and they are nearly two orders larger than the n_2 values reported for thiophene oligomers (Hein, 1994). The $\chi^{(3)}$ values of PThOxad 1a-c are comparable with 5×10^{-12} esu, reported for poly(3-dodecyloxymethylthiophene) (PDTh) (Bredas & Chance, 1989). The values of n_2 and $\chi^{(3)}$ estimated for the polymers PThOxad 1a-c are summarized in Table 2. These values were found to be consistent in all the trials with a maximum error of <10%.

Based on the strong reverse saturable absorption observed for PThOxad 1a-c, good optical limiting action can be expected from these polymers. In general, optical limiters have been utilized in a variety of circumstances where a decreasing transmission with increasing excitation is desirable. However, one of the most important applications is eye and sensor protection in optical systems (Tutt & Boggess, 1993). Fig. 17 demonstrates the optical

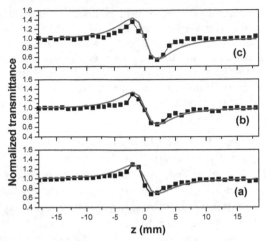

Fig. 16. Pure nonlinear refraction curve obtained for PThOxad 1a (a), PThOxad 1b (b) and PThOxad 1c (c). Solid lines are fit of data to equation (3) with $\Delta\Phi_0 = 1.5$ for PThOxad 1a, $\Delta\Phi_0 = 1.7$ for PThOxad 1b and $\Delta\Phi_0 = 2.2$ for PThOxad 1c.

Polymer	n_o	Z – scan		DFWM			
		n_2 ($\times 10^{-10}$ esu)	β (cm/GW)	Re $\chi^{(3)}$ ($\times 10^{-12}$ esu)	Im $\chi^{(3)}$ ($\times 10^{-12}$ esu)	$\chi^{(3)}$ ($\times 10^{-12}$ esu)	F ($\times 10^{-11}$ esu.cm)
PThOxad 1a	1.422	-1.942	73.8	-2.086	1.139	2.055	4.21
PThOxad 1b	1.422	-2.20	53.0	-2.366	0.818	2.39	3.84
PThOxad 1c	1.415	-2.836	63.6	-3.01	0.968	2.895	4.26

Table 2. Determined values of linear and nonlinear optical parameters for polymers under study.

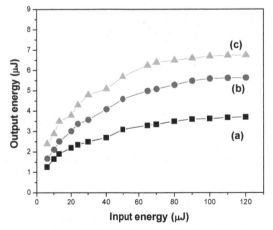

Fig. 17. Optical limiting behavior of PThOxad1a (a), PThOxad1b (b) and PThOxad1c (c), dissolved in DMF solution (1×10^{-5} mol/L).

limiting behavior of PThOxad 1a-c. The optical limiting was recorded by using the open aperture z-scan set up; the output energy transmitted by the sample was measured while the sample was kept fixed at the focus of the lens and the input energy was varied. For incident pule energies up to 20 µJ, the output was linearly increasing with the input. However, for energies more than around 20 µJ, optical limiting of pulses was observed in all the polymers. Although both nonlinear absorption and scattering can contribute to the optical limiting, no significant scattering from the samples was observed during the experiment within the energy limit used.

Concentration dependence of NLO properties can be analyzed to extract information on the NLO properties of the solute. The concentration of polymer in DMF solution was varied and the z-scan experiments were repeated on solutions at each concentration to study the variation of nonlinear response. Fig. 18 shows the dependence of nonlinear absorption (β) on the concentration of PThOxad 1a-c in solution. From 1×10^{-5} to 0.25×10^{-5} mol/L, β decreased linearly with the concentration, however, it was found to decrease rapidly below 0.25×10^{-5} mol/L. The nonlinear absorption as well as the nonlinear refraction decreased as the concentration of polymers in the solution decreased from 1×10^{-5} mol/L to 1.25×10^{-6} mol/L. Similarly the optical limiting behavior of the polymers was found to be dependent on the concentration. The optical limiting behaviors of PThOxad 1a-c were studied at different concentrations (E.g., fig. 19). Both the limiting threshold as well as the clamping level was found to vary the concentration. While the clamping level was found to be increasing with decreasing concentration, the optical limiting threshold was found to be increasing with decreasing concentration of the polymer in the DMF solution.

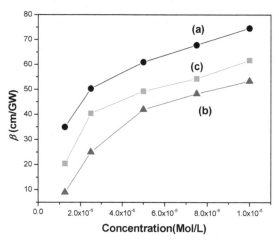

Fig. 18. Concentration dependence of β for polymers PThOxad 1a (a), PThOxad 1b (b) and PThOxad 1c (c).

The third-order NLO properties of two more series of polymers, PThOxad 2a-c and PThOxad 3a-c were also studied using the z-scan technique. The z-scan experiments were performed at pulse energy of 20 µJ which corresponds to a peak irradiance of 0.44 GW/cm². The sign of the nonlinear refractive index of these copolymers was also found to be negative

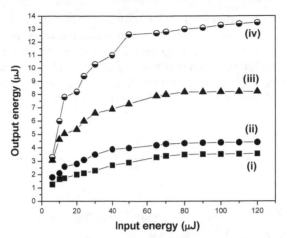

Fig. 19. Optical limiting behaviors of PThOxad 1a-c at different concentrations: (i) 1x10^{-5} Mol/L; (ii) 0.5xX10^{-5} Mol/L; (iii) 0.25x10^{-5} Mol/L: (iv) 0.125x10^{-5} Mol/L.

at 532 nm. The values of n_2 were −1.290x10^{-10}, −1.220x10^{-10} and −1.354x10^{-10} esu for PThOxad 2a-c, respectively. For PThOxad 3a-c, the values of n_2 were −1.121x10^{-10}, −1.212 x10^{-10} and −1.621 x10^{-10} esu, respectively. All the polymers exhibited strong reverse saturable absorption and very good optical limiting properties at 532 nm. The values of two-photon absorption coefficient (β) were 26, 24 and 32 cm/GW for PThOxad 2a-c, respectively. The corresponding values for PThOxad 3a-c were 20, 24 and 36 cm/GW, respectively. The magnitude of the $\chi^{(3)}$ was found to be of the order of 10^{-12} esu for all the polymers. The $\chi^{(3)}$ values were −1.38x10^{-12}, −1.315x10^{-12} esu for PThOxad 2a-c, respectively. For PThOxad 3a-c the corresponding values were −1.08x10^{-12}, −1.186x10^{-12} and −1.564 x10^{-12} esu, respectively.

Variation of the DFWM signal as a function of the pump intensity PThOxad 1c is shown in Fig. 20. The intensity dependence of the amplitude of the DFWM signal in other polymers was found to follow the similar pattern shown in the figure. The signal strength was proportional to the cubic power of the input intensity as given by the equation,

$$I(\omega)\alpha\left(\frac{\omega}{2\varepsilon_0cn^2}\right)^2\left|\chi^{(3)}\right|^2 l^2\, I_0^3(\omega),\qquad(4)$$

where $I(\omega)$ is the DFWM signal intensity, $I_0(\omega)$ is the pump intensity, l is the optical pathlength, and n is the refractive index of the medium. $\chi^{(3)}$ can be calculated from the equation:

$$\chi^{(3)}=\chi^{(3)}_{ref}\left[\frac{I/I_0^3}{(I/I_0^3)_{ref}}\right]^{1/2}\left[\frac{n}{n_{ref}}\right]^2\frac{l_{ref}}{l}\left(\frac{\alpha l}{(1-e^{-\alpha l})e^{-\frac{\alpha l}{2}}}\right),\qquad(5)$$

where the subscript 'ref' refers to the standard reference CS$_2$ under identical conditions, and $\chi^{(3)}_{ref}$ is taken to be 4.0X10^{-13} esu (Philip et al., 1999; Shrik et al., 1992). The figure of merit F was calculated using the equation $\chi^{(3)}/\alpha$. F is a measure of the nonlinear response that can be achieved for a given absorption loss, and is useful in comparing nonlinear materials in the region of absorption. The value of F is given in Table 2, which shows that the polymers possess good F values. It can be noted that the largest $\chi^{(3)}$ and F values have been measured for the polymer attached with highest electron donor among the copolymers. The value of $\chi^{(3)}$ measured by DFWM technique very well matches with the value of $\chi^{(3)}$ obtained by z-scan technique.

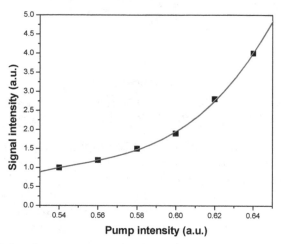

Fig. 20. Variation of the phase conjugate signal with the pump intensity PThOxad 1c. The solid line is a cubic fit of data.

4.2 Structure – NLO property relationship

Polymers PThOxad 1a-c contain alternating electron donating and electron withdrawing groups in their chain. The lengths of the alkoxy groups (-OR) at 3- and 4- positions of the thiophene rings play an important role in the third-order nonlinear response of the polymers. The alkoxy groups present in polymers PThOxad 1a-c are hexyloxy ($-OC_6H_{13}$), octyloxy ($-OC_8H_{17}$) and decyloxy ($-OC_{10}H_{21}$) respectively. The $\chi^{(3)}$ values of the polymers are found to be increasing from PThOxad 1a to PThOxad 1c. This can be attributed to the increase in the electron donating abilities of the alkoxy groups with increase in chain length. That is, the electron donating ability of alkoxy groups is in the order $-OC_6H_{13} < -OC_8H_{17} < -OC_{10}H_{21}$. Hence polymer PThOxad 1c, containing the longest alkoxy groups ($-OC_{10}H_{21}$) at 3- and 4- positions of the thiophene ring, shows the highest nonlinear response among the polymers. Therefore, the enhancement in third-order nonlinear response is attributed to the increased π-electron delocalization in the polymers. A similar variation has been observed in the nonlinear responses of polymers, PThOxad 2a-c. The observed nonlinear response of the polymers are in the order PThOxad 2a < PThOxad 2b < PThOxad 2c. Among three series of polymers (PThOxad 1a-c, PThOxad 2a-c and PThOxad 3a-c), polymers of PThOxad 1 series showed higher values of n_2 and $\chi^{(3)}$. The high nonlinear response of these polymers can be attributed to an increase in the effective conjugation length of the repeating unit due to the presence of vinylene double bonds of 1,4-divinylbenzene unit in the main chain. Compared with PThOxad 3a-c, polymers PThOxad 2a-c show higher nonlinear response due to the presence of stronger electron donating 3,4-ethylenedioxythiophene units. These results suggest that the nonlinear optical properties of the polymers can be tuned by the structural design which involves introduction of alternating donor and acceptor groups along the polymer chain.

4.3 Polymer/TiO₂ nanocomposites

The open aperture z-scan curves obtained for the polymer (PThOxad 3c) film and the PThOxad 3c/TiO$_2$ composite films are shown in figs. 21 and 22 respectively. All the films show a strong optical limiting behavior. The effect is quite strong because the normalized transmission gets decreased to values like 0.1. In polymer composite systems under resonant excitation conditions, an optical limiting behavior can be attributed to effects such as excited state absorption (excited singlet and/or triplet absorption), two- or three-photon absorption (2PA, 3PA), self-focusing/defocusing, thermal blooming, and induced thermal scattering. Of these, 2PA, 3PA, and self focusing/defocusing are electronic nonlinearities that require high laser intensities usually available only from pulsed picosecond or femtosecond lasers. In the present case the possibilities are therefore that of excited state absorption, thermal blooming and induced thermal scattering. We did not visually observe any induced scattering, and the numerical aperture of the detector was large enough to accommodate the transmitted beam fully even if moderate thermal blooming were to happen. Therefore the cause of the observed optical limiting turns out to be excited state absorption. It is possible to model excited state absorption as an "effective" 2PA or 3PA for numerical convenience, and when we tried to fit the experimental data to standard nonlinear transmission equations accordingly, the existence of a relatively weaker saturable absorption also came to light. Therefore an effective intensity-dependent nonlinear absorption coefficient of the form,

$$\alpha(I) = \frac{\alpha_0}{1 + (I/I_s)} + \beta I \tag{6}$$

can be considered, where α_0 is the unsaturated linear absorption coefficient at the wavelength of excitation, I is the input laser intensity, and I_s is the saturation intensity (intensity at which the linear absorption drops to half its original value). $\beta I = \sigma N$ is the excited state absorption (ESA) coefficient, where σ is the ESA cross section and $N(I)$ is the intensity-dependent excited state population density. For calculating the transmitted intensity for a given input intensity, the propagation equation,

$$\frac{dI}{dz'} = -\left[\left(\alpha_0 / \left(1 + \frac{I}{I_s}\right)\right) + \beta I\right] I \tag{7}$$

was numerically solved. Here z' indicates the propagation distance within the sample. By determining the best-fit curves for the experimental data, the nonlinear parameters could be calculated.

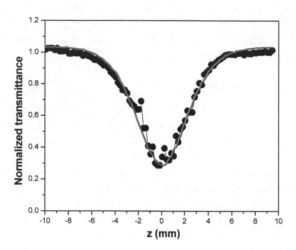

Fig. 21. Open aperture z-scan of polymer film having a linear transmission of 59 % at 532 nm. The laser pulse energy is 90 microJoules. Circles are data points while the solid curve is a numerical fit according to equation (6).

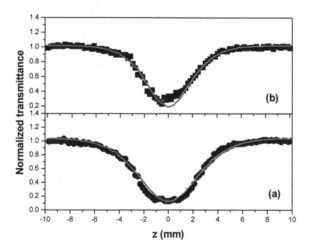

Fig. 22. Open aperture z-scan of nanocomposite films having a linear transmission of 51% at 532 nm. The laser pulse energy is a) 100 microJoules, b) 75 microJoules. Circles are data points while the solid curve is a numerical fit according to equation (6).

For PThOxad 3c film, β is found to be 1×10^{-7} m/W, while for the nanocomposite film it is 2×10^{-7} m/W. Obviously there is an enhancement of nonlinearity in PThOxad 3c/TiO$_2$

nanocomposite film compared to the pure PThOxad 3c film. This is not substantial though, the reason being that the polymer films themselves are highly nonlinear in nature. From a device point of view both PThOxad 3c and PThOxad 3c/TiO$_2$ nanocomposite films are equally useful because the optical limiting efficiency exhibited by them is high. To put the above β values in perspective, the values obtained in similar systems under similar excitation conditions are, 6.0 x 10^{-8} m/W in p-(N,N-dimethylamino)dibenzylideneacetone in PMMA matrix (Kiran et al., 2008), 10^{-7} to 10^{-9} m/W in Au:Ag-PVA nanocomposite films and 6.8 x 10^{-7} m/W in a ZnO/PMMA nanocomposite (Sreeja et al., 2010). Obviously, the present films are potentially suited for fabricating optical limiters, which can protect sensitive light detectors and also human/animal eyes from accidental exposure to high levels of optical radiation, while maintaining normal transparency for safe low level inputs.

5. Conclusions

The third-order nonlinear optical properties of three series of Donor-Acceptor polymers containing oxadiazole and substituted thiophene units and of a polymer/TiO$_2$ nanocomposite film have been investigated by using z-scan and DFWM techniques in the nanosecond domain. The results indicated that a nonlinear refractive index of the order of 10^{-10} esu can be readily obtained in these D-A polymers. The magnitude of $\chi^{(3)}$ is found to be of the order of 10^{-12} esu for all the polymers. The nonlinear absorption is found to be originating from the reverse saturable absorption. The polymers exhibit good optical limiting properties at 532 nm. The results showed that the polymer PThOxad 1c, containing the longest alkoxy group, exhibits highest nonlinearity among the three series of polymers and it may be a potential candidate for optical limiting, optical switching and other fast photonic applications. The dependence of NLO parameters on the length of the alkoxy substituents present in the polymers indicates that the nonlinearity in these polymers originates from the electronic effects. The nonlinear absorption, nonlinear refraction and optical limiting behavior of the polymers increase with increasing sample concentration in solution. In PThOxad 3c/TiO$_2$ nanocomposite, incorporation of TiO$_2$ nanoparticles into the polymer matrix is found to improve the thermal stability of the polymer. A red shift in the absorption spectra and a blue shift in the emission spectra are observed for PThOxad 3c/TiO$_2$ nanocomposite film compared to those of PThOxad 3c film. The composite film showed a strong optical limiting behavior, and the incorporation of TiO$_2$ marginally enhanced the nonlinear absorption coefficient value of the polymer. The present study reveals that the third-order nonlinear optical properties of conjugated polymers can be enhanced by increasing the π-electron delocalization along the polymer chain through a proper structural modification. Further, the results show that the nonlinear effects in a conjugated polymer can be enhanced by incorporating TiO$_2$ nanoparticles in to the polymer matrix.

6. Acknowledgements

We thank Dr. Reji Philip, Raman Research Centre Bangalore, India for useful suggestions and for providing facility for the NLO experiments.

7. References

Boyd, R.W.; Gehr, R.J.; Fischer, G.L. & Sipe, J.E. (1996). Nonlinear Optical Properties of Nanocomposite Materials. *Pure and Applied Optics*, Vol.5, No.5, (September 1996), pp. 505-512, ISSN 1361-6617.

Bredas, J.L. & Chance, R.R. (1989). *Conjugated Polymeric Materials Opportunities in Electronics, Optoelectronics and Molecular Electronics*, ISBN 13-9780792307518, Kluwer Academic publishers, USA.

Brocks, G. & Tol, J. (1996). Small Band Gap Semiconducting Polymers Made from Dye Molecules: Polysquaraines. *Journal of Physical Chemistry*, Vol.100, No.5, (February 1996), pp. 1838-1848, ISSN 1520-5207.

Cassano, T.; Tommasi, R.; Babudri, F.; Cardone, A.; Farinola, G.M. & Naso, F. (2002). High Third-Order Nonlinear Optical Susceptibility in New Fluorinated Poly(p-phenylenevinylene) Copolymers Measured with the Z-scan Technique. *Optical Letters*, Vol.27, No.24, (December 2002), pp. 2176-2178, ISSN 1539-4794.

Chen, Q.; Sargent, E.H.; Leclerc, N. & Attias, A.J. (2003). Wavelength Dependence and Figures of Merit of Ultrafast Third-Order Optical Nonlinearity of a Conjugated 3,3'-Bipyridine Derivative. *Applied Optics*, Vol.42, No.36, (December 2003), pp. 7235-7241, ISSN 2155-3165.

Chen, X.; Tao, J.; Zou, G.; Zhang, Q. & Wang, P. (2010). Nonlinear Optical Properties of Nanometer-Size Silver Composite Azobenzene Containing Polydiacetylene Film. *Applied Physics A*, Vol.100, No.1, (February 2010), pp. 223-230, ISSN 1432-0630.

Couris, S.; Koudoumas, E.; Ruth, A.A. & Leach, S. (1995). Concentration and Wavelength Dependence of the Effective Third-Order Susceptibility and Optical Limiting of C_{60} in Toluene Solution. *Journal of Physics B: Atomic, Molecular and Optical Physics*, Vol.28, No.20, (October 1995), pp. 4537-4554, ISSN 1361-6455.

Davey, A.P.; Elliott, S.; Connor, O.O. & Blau, W. (1995). New Rigid Backbone Conjugated Organic Polymers with Large Fluorescence Quantum Yields. *Journal of the Chemical Society, Chemical Communications*, No.14, pp. 1433-1434, (January 1995), ISSN 0022-4936.

Demas, J.N. & Crosby, G.A. (1971). Measurement of Photoluminescence Quantum Yields. *Journal of Physical Chemistry*, Vol.75, No.8, (April 1971), pp. 991-1024, ISSN 1932-7447.

Fatti, N.D. & Vallee, F. (2001). Ultrafast Optical Nonlinear Properties of Metal Nanoparticles. *Applied Physics B*, Vol.73, No.4, (October 2001), pp. 383-390, ISSN 1432-0649.

Gao, Y.; Tonizzo, A.; Walser, A.; Potasek, M. & Dorsinville, R. (2008). Enhanced Optical Nonlinearity of Surfactant-Capped CdS Quantum Dots Embedded in an Optically Transparent Polystyrene Thin Film. *Applied Physics Letters*, Vol.92, No.3, (January 2008), pp. 033106-033109, ISSN 1077-3118.

Gayvoronsky, V.; Galas, A.; Shepelyavyy, E.; Dittrich, Th.; Timoshenko, V.Yu.; Nepijko, S.A.; Brodyn, M.S. & Koch, F. (2005). Giant Nonlinear Optical Response of Nanoporous Anatase Layers. *Applied Physics B*, Vol.80, No.1, (November 2005), pp. 97-100, ISSN 1432-0649.

Guo, S.Li.; Xu, L.; Wang, H.T.; You, X.Z. & Ming, N.B. (2003). Investigation of Optical Nonlinearities in Pd(po)$_2$ by Z-scan Technique. *Optik - International Journal for Light and Electron Optics*, Vol.114, No.2, (November 2003), pp. 58-62, ISSN 0030-4026.

Havinga, E.E.; Hoeve, W. & Wynberg, H. (1993). Alternate Donor-Acceptor Small Band Gap Semiconducting Polymers; Polysquaraines and Polycroconaines. *Synthetic Metals*, Vol.55, No.1, (March 1993), pp. 299-306, ISSN 03796779.

Hegde,; P.K. Vasudeva Adhikari,; A. Manjunatha, M.G.; Suchand Sandeep, C.S. & Reji Philip (2009). Synthesis and Nonlinear Optical characterization of New poly{2,2'-(3,4-didodecyloxythiophene-2,5-diyl)bis[5-(2-thienyl)-1,3,4-oxadiazole]}. *Synthetic Metals*, Vol.159, No.11, (June 2009), pp. 1099–1105, ISSN 0379-6779.

Hein, J.; Bergner, H.; Lenzner, M. & Rentsch, S. (1994). Determination of Real and Imaginary Part of $\chi^{(3)}$ of Thiophene Oligomers using the Z-scan Technique. *Chemical Physics*, Vol.179, No.3, (February 1994), pp. 543-548, ISSN 0301-0104.

Henari, F.Z.; Blau, W.J.; Milgrom, L.R.; Yahioglu, G.; Philips, D. & Lacey, J.A. (1997). Third-Order Optical Non-Linearity in Zn(II) Complexes of 5,10,15,20-tetraarylethynyl-Substituted Porphyrins. *Chemical Physics Letters*, Vol.267, No.3-4, (March 1997), pp. 229-233, ISSN 0009-2614.

Hsieh, S.N.; Wen, T.C. & Guo, T.F. (2007). Polymer/Gold Nanoparticles Light Emitting Diodes Utilizing High Work Function Metal Cathodes. *Materials Chemistry and Physics*, Vol.101, No.2-3, (February 2007), pp. 383-386, ISSN 0254-0584.

Hu X.Y.; Jiang, P.; Ding, C.Y.; Yang, H. & Gong, Q.H. (2008). Picosecond and Low-Power all-Optical Switching based on an Organic Photonic-Bandgap Microcavity. *Nature Photonics*, Vol.2, No.3, (March 2008), pp. 185-189, ISSN 1749-4885.

Hu, X.; Zhang, J.; Yang, H. & Gong, Q. (2009). Tunable Time Response of the Nonlinearity of Nanocomposites by Doping Semiconductor Quantum Dots. *Optics Express*, Vol.17, No. 21, (October 2009), PP. 18858-18865, ISSN 1094-4087.

Kamanina, N.V. & Plekhanov, A.I. (2002). Mechanisms of Optical Limiting in Fullerene-Doped π-Conjugated Organic Structures Demonstrated with Polyimide and COANP Molecules. *Optics and Spectroscopy*, Vol.93, No.3, (September 2002), pp. 408-415, ISSN 1562-6911.

Kamanina, N.V. (1999). Reverse Saturable Absorption in Fullerene-Containing Polyimides, Applicability of the Förster Model. *Optical Communication*, Vol.162, No.4-6, (April 1999), pp. 228-232, ISSN 0030-4018.

Kamanina, N.V. (2001). Peculiarities of Optical Limiting Effect in π-Conjugated Organic Systems based on 2-cyclooctylamino-5-nitropyridinedoped with C$_{70}$. *Journal of Optics A Pure and Applied Optics*, Vol.3, No.5, (July 2001), pp. 321-325, ISSN 1464-4258.

Kamanina, N.V.; Reshak, A.H.; Vasilyev, P.Ya.; Vangonen, A.I.; Studeonov, V.I.; Usanov, Yu.E.; Ebothe, J.; Gondek, E.; Wojcik, W. & Danel, A. (2009). Nonlinear Absorption of Fullerene and Nanotubes Doped Liquid Crystal Systems. *Physica E*, Vol.41, No.3, (January 2009), pp. 391-394, ISSN 13869477.

Kamanina, N.V.; Vasilyev, P.Ya.; Vangonen, A.I.; Studeonov, V.I.; Usanov, Yu.E.; Kajzar, F. & Attias, A.J. (2008). Photophysics of Organic Structures Doped with Nanoobjects:

Optical Limiting, Switching and Laser Strength. *Moleculer Crystals and Liquid crystals*, Vol.485, No.1, (April 2008), pp. 945-954, ISSN 1563-5287.

Karthikeyan, B.; Anija, M. & Reji Philip, (2006). In Situ Synthesis and Nonlinear Optical Properties of Au:Ag Nanocomposite Polymer films. *Applied Physics Letters*, Vol.88, No.5, (January 2006), pp. 053104-053107, ISSN 1077-3118.

Kiran, A.J.; Rai, N.S.; Udayakumar, D.; Chandrasekharan, K.; Kalluraya, B.; Reji Philip, Shashikala, H.D. & Adhikari, A.V. (2008). Nonlinear Optical properties of p-(N,N-dimethylamino) dibenzylidene -acetone Doped Polymer. *Material Research Bulletin*, Vol.43, No.3-4, (March 2008), pp. 707-713, ISSN 0025-5408.

Kiran, A.J.; Udayakumar, D.; Chandrasekharan, K.; Adhikari, A.V. & Shashikala, H.D. (2006). Z-scan and Degenerate Four Wave Mixing Studies on Newly Synthesized Copolymers Containing Alternating Substituted Thiophene and 1,3,4-oxadiazole units. *Journal of physics B: Atomic molecular and optical physics*, Vol.39. No.18, (September 2006), pp. 3747-3756, ISSN 0953-4075

Kishino, S.; Ueno, Y.; Ochiai, K.; Rikukawa, M.; Sanui, K.; Kobayashi, T.; Kunugita, H.; & Ema, K. (1998). Estimate of the effective conjugation length of polythiophene from its $|\chi^{(3)}(\omega;\omega,\omega,-\omega)|$ spectrum at excitonic resonance. *Physical Review B*, Vol. 58, No. 20, PP. R13430-R13433, (November 1998) ISSN 1943-2879.

Liu, X.; Guo, S.; Wang, H. & Hou, L. (2001). Theoretical Study on the Closed-Aperture Z-scan Curves in the Materials with Nonlinear Refraction and Strong Nonlinear Absorption. *Optics Communications*, Vol.197, No.4-6, (October 2001), pp. 431-437, ISSN 0030-4018.

Munn, R.W. & Ironside, C.N. (1993). *Principles and Applications of Nonlinear Optical Materials*, ISBN 10-0751400858, Routledge Chapman Hall.

Neeves, A.E. & Birnboim, M.H. (1988). Composite Structures for the Enhancement of Nonlinear Optical Materials. *Optics Letters*, Vol.13, No.12, (December 1988), pp. 1087-1089, ISSN 1539-4794.

Nisoli, M.; Cybo-Ottone, A.; De Silvestri, S.; Magni, V.; Tubino, R.; Botta, C. & Musco, A. (1993). Femtosecond Transient Absorption Saturation in Poly(alkyl-thiophene-vinylene)s. *Physical Review B*, Vol.47, No.16, (April 1993), pp. 10881-10884, ISSN 1943-2879.

Philip, R.; Ravikanth, M. & Ravindrakumar, G. (1999). Studies of Third Order Optical Nonlinearity in Iron (III) Phthalocyanine μ-oxo Dimers Using Picosecond Four-wave Mixing. *Optics Communications*, Vol.165, No.1-3, (July 1999), pp. 91-97, ISSN 0030-4018.

Porel, S.; Venkataram, N.; Rao, D.N. & Radhakrishnan, T.P. (2007). Optical Power Limiting in the Femtosecond Regime by Silver Nanoparticle–Embedded Polymer Film. *Journal Applied Physics*, Vol.102, No.3, (August 2007), pp. 033107-033113, ISSN 1077-3118.

Prasad, P.N. & Williams, D.J. (1992). *Introduction to Nonlinear Optical Effects in Molecules and Polymers*, ISBN 978-0-471-51562-3, John Wiley & Sons, New York.

Reji Philip; Ravindrakumar, G.; Sandhyarani, N. & Pradeep, T. (2000). Picosecond Optical Nonlinearity in Monolayer-Protected Gold, Silver, and Gold-Silver Alloy

Nanoclusters. *Physical Review B*, Vol.62, No.19, (November 2000), pp. 13160-13166, ISSN 1943-2879.

Sezer, A.; Gurudas, U.; Collins, B.; Mckinlay, A. & Bubb, D.M. (2009). Nonlinear Optical Properties of Conducting Polyaniline and Polyaniline–Ag Composite Thin Films. *Chemical Physics Letters*, Vol.477, No.1-3, (July 2009), ISSN 0009-2614.

Sheik Bahae, M.; Said, A.A.; Wei, T.H.; Hagan, D.J. & Stryland, E.W.Van. (1990). Sensitive Measurement of Optical Nonlinearities using a Single Beam. *IEEE Journal of Quantum Electronics*, Vol.26, No.4, (April 1990), pp. 760-769, ISSN 0018-9197.

Shrik, J.S.; Lindle, J.R.; Bartoli, F.J. & Boyle, M.E. (1992). Third-Order Optical Nonlinearities of bis(phthalocyanines). *Journal of Physical Chemistry*, Vol.96, No.14, (July 1992), pp. 5847-5852, ISSN 1932-7447.

Sreeja, R.; John, J.; Aneesh, P.M. & Jayaraj, M.K. (2010). Linear and Nonlinear Optical Properties of Luminescent ZnO Nanoparticles Embedded in PMMA Matrix. *Optics Communications*, Vol.283, No.14-15, (July 2010), pp. 2908-2913, ISSN 0030-4018.

Sutherland, R.L. (1996). *Hand Book of Nonlinear Optics*, ISBN 0-8247-9651-9, Marcel Dekker Inc., New York.

Takele, H.; Greve, H.; Pochstein, C.; Zaporojtchenko, V. & Faupel, F. (2006). Plasmonic Properties of Ag Nanoclusters in Various Polymer Matrices. *Nanotechnology*, Vol.17, No.14, (July 2006), pp. 3499-3504, ISSN 1361-6528.

Tutt, L.W. & Boggess T.F. (1993). A Review of Optical Limiting Mechanisms and Devices using Organics, Fullerenes, Semiconductors and other Materials. *Progress in Quantum Electronics*, Vol.17, No.4, (August 1993), pp. 299-338, ISSN 0079-6727.

Udayakumar, D.; John Kiran, A.; Adhikari, A.V.; Chandrasekharan, K. & Shashikala, H.D. (2007). Synthesis and Nonlinear Optical Characterization of Copolymers Containing Alternating 3,4-dialkoxythiophene and (1,3,4-oxadiazolyl)benzene units. *Journal of Applied Polymer Science*, Vol. 106, No.5, (December 2007), pp. 3033–3039, ISSN 1097-4628.

Udayakumar, D.; John Kiran, A.; Vasudeva Adhikari, A.; Chandrasekharan, K.; Umesh, G. & Shashikala H.D. (2006). Third-Order Nonlinear Optical Studies of Newly Synthesized polyoxadiazoles Containing 3,4-dialkoxythiophenes. *Chemical Physics*, Vol.331, No.1, (December 2006), pp. 125-130, ISSN 0301-0104.

Venkatram, N.; Rao, D.N. & Akundi, M.A. (2005). Nonlinear Absorption, Scattering and Optical Limiting Studies of CdS nanoparticles. *Optics Express*, Vol.13, No.3, (February 2005), PP. 867-872, ISSN 1094-4087.

Voisin, C. ; Del Fatti, N.; Christofilos, D. & Vall´ee F. (2001). Ultrafast Electron Dynamics and Optical Nonlinearities in Metal Nanoparticles. *Journal of Physical Chemistry B*, Vol.105, No.12, (March 2001), pp. 2264-2280, ISSN 1089-5647.

Yang, P.; Xu, J.; Ballato, J.; Schwartz, R.W. & Carroll, D.L. (2002). Optical Limiting in $SrBi_2Ta_2O_9$ and $PbZr_xTi_{1-x}O_3$ Ferroelectric Thin Films. *Applied Physics Letters*, Vol.80. No.18, (April 2002), pp. 3394-3397, ISSN 1077-3118.

Yang, S.H.; Rendu, P.L.; Nguyen, T.P. & Hsu, C.S. (2007). Fabrication of MEH-PPV/SiO_2 and MEH-PPV/TiO_2 Nanocomposites with Enhanced Luminescent Stabilities. *Reviews on Advanced Materials Science*, Vol.15, No.2,

Zhu, Y.; Xu, S.; Jiang, L.; Pan, K. & Dan, Y. (2008). Synthesis and Characterization of Polythiophene/Titanium dioxide Composites. *Reactive and Functional Polymers*, Vol.68, No.10, (October 2008), pp. 1492-1498, ISSN 1381-5148.

Zyss, J. (1994). *Molecular Nonlinear Optics, Materials, Physics and Devices*, ISBN 978-0-12-784450-3, Academic Press.

6

Reflection and Transmission of a Plane TE-Wave at a Lossy, Saturating, Nonlinear Dielectric Film

Valery Serov[1] and Hans Werner Schürmann[2]
[1]University of Oulu
[2]University of Osnabrück
[1]Finland
[2]Germany

1. Introduction

Reflection and transmission of transverse-electric (TE) electromagnetic waves at a single nonlinear homogeneous, isotropic, nonmagnetic layer situated between two homogeneous, semi-infinite media has been the subject of intense theoretical and experimental investigations in recent years. In particular, the Kerr-like nonlinear dielectric film has been the focus of a number of studies in nonlinear optics (Chen & Mills, 1987; 1988; Leung, 1985; 1988; Peschel, 1988; Schürmann & Schmoldt, 1993).

Exact analytical solutions have been obtained for the scattering of plane TE-waves at Kerr-nonlinear films (Leung, 1989; Schürmann et al., 2001). As far as exact analytical solutions were considered in these articles absorption was excluded, at most it was treated numerically (Gordillo-Vázquez & Pecharromán, 2003; Schürmann & Schmoldt, 1996; Yuen & Yu, 1997).

As Chen and Mills have pointed out it is a nontrivial extension of the usual scattering theory to include absorption (Chen & Mills, 1988) and it seems (to the best of our knowledge) that the problem was not solved till now. In the following we consider a nonlinear lossy dielectric film with spatially varying saturating permittivity. In Section 2 we reduce Maxwell's equations to a Volterra integral equation (14) for the intensity of the electric field $E(y)$ and give a solution in form of a uniform convergent sequence of iterate functions. Using these solutions we determine the phase function $\vartheta(y)$ of the electric field, and, evaluating the boundary conditions in Section 3, we derive analytical expressions for reflectance, transmittance, absorptance, and phase shifts on reflection and transmission and present some numerical results in Section 4.

It should be emphasized, that the contraction principle (that is used in this work) (Zeidler, 1995) includes the proof of the existence of the exact bounded solution of the problem and additionally yields approximate analytical solutions by iterations. Furthermore, the rate of convergence of the iterative procedure and the error estimate can be evaluated (Zeidler, 1995). Thus this approach is useful for physical applications.

The present approach can be applied to a linear homogeneous, isotropic, nonmagnetic layer with absorption. In this case the problem is reduced to a linear Volterra integral equation that can be solved by iterations without any restrictions. The lossless linear permittivity as well as the Kerr-like permittivity can be treated as particular cases of the approach.

Referring to figure 1 we consider a dielectric film between two linear semi-infinite media (substrate and cladding). All media are assumed to be homogeneous in $x-$ and $z-$ direction, isotropic, and non-magnetic. The film is assumed to be absorbing and characterized by a complex valued permittivity function $\varepsilon_f(y)$.

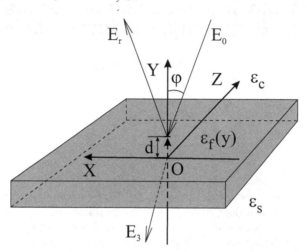

Fig. 1. Configuration considered in this paper. A plane wave is incident to a nonlinear slab (situated between two linear media) to be reflected and transmitted.

A plane wave of frequency ω_0 and intensity E_0^2, with electric vector $\mathbf{E_0}$ parallel to the z-axis (TE) is incident on the film of thickness d. Since the geometry is independent of the z-coordinate and because of the supposed TE-polarization fields are parallel to the z-axis ($\mathbf{E} = (0, 0, E_z)$). We look for solution $\mathbf{E}$ of Maxwell's equations

$$rot\mathbf{H} = -i\omega_0\varepsilon\mathbf{E}$$

$$rot\mathbf{E} = i\omega_0\mu_0\mathbf{H}$$

that satisfy the boundary conditions (continuity of E_z and $\partial E_z/\partial y$ at interfaces $y \equiv 0$ and $y \equiv d$) and where (due to TE-polarization) $\mathbf{H} = (H_x, H_y, 0)$. Due to the requirement of the translational invariance in x-direction and partly satisfying the boundary conditions the fields tentatively are written as ($\hat{z}$ denotes the unit vector in z-direction)

$$\mathbf{E}(x, y, t) = \begin{cases} \hat{z}\left[E_0 e^{i(px - q_c\cdot(y-d) - \omega_0 t)} + \\ \quad E_r e^{i(px + q_c\cdot(y-d) - \omega_0 t)}\right], & y > d, \\ \hat{z}\left[E(y)e^{i(px + \vartheta(y) - \omega_0 t)}\right], & 0 < y < d, \\ \hat{z}\left[E_3 e^{i(px - q_s y - \omega_0 t)}\right], & y < 0, \end{cases} \tag{1}$$

where $E(y)$, $p = \sqrt{\varepsilon_c}k_0 \sin\varphi$, $k_0 = \omega_0\sqrt{\varepsilon_0\mu_0}$, q_c, q_s, and $\vartheta(y)$ are real and $E_r = |E_r|\exp(i\delta_r)$ and $E_3 = |E_3|\exp(i\delta_t)$ are independent of y.

We assume a permittivity $\varepsilon(y)$ of the three layer system modeled by

$$\frac{\varepsilon(y)}{\varepsilon_0} = \begin{cases} \varepsilon_c, \ y > d, \\ \varepsilon_f(y) = \varepsilon_f^0 + \varepsilon_R(y) + i\varepsilon_I(y) + \frac{aE^2(y)}{1+arE^2(y)}, \ 0 < y < d, \\ \varepsilon_s, \ y < 0, \end{cases} \tag{2}$$

with real constants $\varepsilon_c, \varepsilon_s, \varepsilon_f^0, a \geq 0, r \geq 0$ and real-valued continuously differentiable functions $\varepsilon_R(y), \varepsilon_I(y)$ on $[0,d]$. A particular nonlinearity in (2) of cubic type ($r = 0$) can be met in the context of a Kerr-like nonlinear dielectric film, while the case when $r > 0$ corresponds to the saturation model in optics (see (Bang et al., 2002; Berge et al., 2003; Dreischuh et al., 1999; Kartashov et al., 2003)).

The problem to be solved is to find a solution of Maxwell's equations subject to (1) and (2). With respect to the physical significance of (1) and (2) some remarks may be appropriate. Though ansatz (1) widely has been used previously (Chen & Mills, 1987; 1988; Leung, 1985; 1988; Peschel, 1988; Schürmann & Schmoldt, 1993) it should be noted that it is based on the assumption that the time-dependence of the optical response of the nonlinear film is described by one frequency ω_0. Phase matching is, e.g., assumed to be absent so that small amplitudes of higher harmonics can be neglected. The permittivity function (2) also represents an approximation. The dipole moment per unite volume and hence the permittivity is not simply controlled by the instant value of the electric (macroscopic) field at the point (x,y,z), due to the time lag of the medium's response. Further more the response is nonlocal in space. - The model permittivity (2) does not incorporate these features. Nevertheless, experimental observations (cf., e.g. (Peschel, 1988)) indicate that (2) has physical significance (with $\varepsilon_R = \varepsilon_I = r = 0$). Finally, Maxwell's equations, even for an isotropic material, imply that all field components are coupled if the permittivity is nonlinear. The decomposition into TE-and TM-polarization is an assumption motivated by mathematical simplicity. To apply the results below to experiments it is necessary to make sure that TE-polarization is maintained.

2. Nonlinear Volterra integral equation

By inserting (1) and (2) into Maxwell's equations we obtain the nonlinear Helmholtz equations, valid in each of the three media ($j = s, f, c$),

$$\frac{\partial^2 \tilde{E}_j(x,y)}{\partial x^2} + \frac{\partial^2 \tilde{E}_j(x,y)}{\partial y^2} + k_0^2 \frac{\varepsilon(y)}{\varepsilon_0} \tilde{E}_j(x,y) = 0, \quad j = s, f, c, \tag{3}$$

where $\tilde{E}_j(x,y)$ denotes the time-independent part of $\mathbf{E}(x,y,t)$.

Scaling x, y, z, p, q_c, q_s by k_0 and using the definition of $\frac{\varepsilon}{\varepsilon_0}$ in equation (2), equation (3) reads

$$\frac{\partial^2 \tilde{E}_j(x,y)}{\partial x^2} + \frac{\partial^2 \tilde{E}_j(x,y)}{\partial y^2} + \varepsilon_j(y)\tilde{E}_j(x,y) = 0, \quad j = s, f, c, \tag{4}$$

where the same symbols have been used for unscaled and scaled quantities. Using ansatz (1) in equation (4) we get for the semi-infinite media

$$q_j^2 = \varepsilon_j - p^2, \quad j = s, c. \tag{5}$$

For the film $(j = f)$, we obtain, omitting tildes,

$$\frac{d^2E(y)}{dy^2} - E(y)\left(\frac{d\vartheta(y)}{dy}\right)^2 + \left(\varepsilon_f^0 + \varepsilon_R(y) - p^2 + \frac{aE^2(y)}{1+arE^2(y)}\right)E(y) = 0 \tag{6}$$

and

$$E(y)\frac{d^2\vartheta(y)}{dy^2} + 2\frac{d\vartheta(y)}{dy}\frac{dE(y)}{dy} + \varepsilon_I(y)E(y) = 0. \tag{7}$$

Equation (7) can be integrated leading to

$$E^2(y)\frac{d\vartheta(y)}{dy} = c_1 - \int_0^y \varepsilon_I(\tau)E^2(\tau)d\tau, \tag{8}$$

where c_1 is a constant that is determined by means of the boundary conditions:

$$c_1 = E^2(0)\frac{d\vartheta(0)}{dy} = -q_s E^2(0) \tag{9}$$

Insertion of $d\vartheta(y)/dy$ according to equation (3) leads to

$$\frac{d^2E(y)}{dy^2} + (q_f^2(y) - p^2)E(y) + \frac{aE^3(y)}{1+arE^2(y)} - \frac{\left(c_1 - \int\limits_0^y \varepsilon_I(t)E^2(t)dt\right)^2}{E^3(y)} = 0, \tag{10}$$

with

$$q_f^2(y) = \varepsilon_f^0 + \varepsilon_R(y). \tag{11}$$

As for real permittivity, real q_s (transmission) implies $c_1 \neq 0$.

Setting $I(y) = aE^2(y), a \neq 0$, multiplying equation (10) by $4E^3(y)$, and differentiating the result with respect to y we obtain

$$\frac{d^3I(y)}{dy^3} + 4\frac{d\left((q_f^2(y) - p^2)I(y)\right)}{dy} = 2\frac{d(q_f^2(y))}{dy}I(y)$$

$$- \frac{2I(y)\frac{dI(y)}{dy}(3+2rI(y))}{(1+rI(y))^2}$$

$$- 4\varepsilon_I(y)(ac_1 - \int\limits_0^y \varepsilon_I(t)I(t)dt). \tag{12}$$

Equation (12) can be integrated with respect to $I(y)$ to yield

$$\frac{d^2I(y)}{dy^2} + 4\kappa^2I(y) = -4\varepsilon_R(y)I(y) + 2\int_0^y \frac{d\varepsilon_R(t)}{dt}I(t)dt$$

$$- \frac{2}{r^2}\left(2rI(y) + \frac{1}{1+rI(y)} - \ln(1+rI(y))\right)$$

$$+ 4\int_0^y \varepsilon_I(t)\left(\int_0^t \varepsilon_I(z)I(z)dz\right)dt - 4ac_1\int_0^y \varepsilon_I(t)dt + c_2, \tag{13}$$

where $\kappa^2 = \varepsilon_f^0 - p^2$ and c_2 is a constant of integration.

In the case $a = 0$ (it corresponds to the linear case with absorption in Eq. (2)) we obtain the following analog of (13)

$$\frac{d^2 I(y)}{dy^2} + 4\kappa^2 I(y) = -4\varepsilon_R(y)I(y) + 2\int_0^y \frac{d\varepsilon_R(t)}{dt} I(t)dt$$

$$+4\int_0^y \varepsilon_I(t)\left(\int_0^t \varepsilon_I(z)I(z)dz\right)dt - 4c_1\int_0^y \varepsilon_I(t)dt + c_2, \tag{14}$$

where $I(y)$ denotes $E^2(y)$. Later on for the case $a = 0$ under $I(y)$ we always understand $E^2(y)$.

The homogeneous equation $d^2 I(y)/dy^2 + 4\kappa^2 I(y) = 0$ which corresponds to Eq. (13) has the solution that satisfies the boundary conditions at $y = 0$

$$\tilde{I}_0(y) = a|E_3|^2 \cos(2\kappa y), \tag{15}$$

so that the general solution of equation (13) reads (Stakgold, 1967)

$$I(y) = \tilde{I}_0(y) + \int_0^y dt \frac{\sin 2\kappa(y - t)}{2\kappa} \cdot$$

$$\left(-4\varepsilon_R(t)I(t) + 2\int_0^t \frac{d\varepsilon_R(\tau)}{d\tau} I(\tau)d\tau\right.$$

$$-\frac{2}{r^2}\left(2rI(t) + \frac{1}{1 + rI(t)} - \ln(1 + rI(t))\right)$$

$$+4\int_0^t \varepsilon_I(\tau)\left(\int_0^\tau \varepsilon_I(z)I(z)dz\right)d\tau - 4ac_1\int_0^t \varepsilon_I(\tau)d\tau + c_2\Bigg), \tag{16}$$

where the constant c_2 must be determined by the boundary conditions.

In the case $a = 0$ the general solution of equation (14) reads

$$I(y) = \tilde{I}_0(y) + \int_0^y dt \frac{\sin 2\kappa(y - t)}{2\kappa} \cdot$$

$$\left(-4\varepsilon_R(t)I(t) + 2\int_0^t \frac{d\varepsilon_R(\tau)}{d\tau} I(\tau)d\tau\right.$$

$$+4\int_0^t \varepsilon_I(\tau)\left(\int_0^\tau \varepsilon_I(z)I(z)dz\right)d\tau - 4c_1\int_0^t \varepsilon_I(\tau)d\tau + c_2\Bigg) \tag{17}$$

with $\tilde{I}_0(y) = |E_3|^2 \cos(2\kappa y)$.

The Volterra equations (16), (17) are equivalent to equation (3) for $0 < y < d$ for $a \neq 0$ and $a = 0$, respectively. According to equations (16) and (17) $I(y)$ and $\tilde{I}_0(y)$ satisfy the boundary conditions at $y = 0$. Evaluating some of the integrals on the right-hand side, equations (16) and (17) can be written as

$$
\begin{aligned}
I(y) = I_0(y) &+ \frac{1}{\kappa^2} \int_0^y \sin^2 \kappa(y-\tau) \frac{d\varepsilon_R(\tau)}{d\tau} I(\tau) d\tau \\
&- \frac{2}{\kappa} \int_0^y \sin 2\kappa(y-\tau) \varepsilon_R(\tau) I(\tau) d\tau \\
&- \frac{2}{r\kappa} \int_0^y \sin 2\kappa(y-\tau) I(\tau) d\tau \\
&- \frac{1}{\kappa r^2} \int_0^y \sin 2\kappa(y-\tau) \frac{1}{1+rI(\tau)} d\tau \\
&+ \frac{1}{\kappa r^2} \int_0^y \sin 2\kappa(y-\tau) \ln(1+rI(\tau)) d\tau \\
+ 4 \int_0^y \varepsilon_I(\tau) &\left(\int_\tau^y \frac{\sin 2\kappa(y-t)}{2\kappa} \left(\int_\tau^t \varepsilon_I(z) dz \right) dt \right) I(\tau) d\tau,
\end{aligned}
\tag{18}
$$

and

$$
\begin{aligned}
I(y) = I_0(y) &+ \frac{1}{\kappa^2} \int_0^y \sin^2 \kappa(y-\tau) \frac{d\varepsilon_R(\tau)}{d\tau} I(\tau) d\tau \\
&- \frac{2}{\kappa} \int_0^y \sin 2\kappa(y-\tau) \varepsilon_R(\tau) I(\tau) d\tau \\
+ 4 \int_0^y \varepsilon_I(\tau) &\left(\int_\tau^y \frac{\sin 2\kappa(y-t)}{2\kappa} \left(\int_\tau^t \varepsilon_I(z) dz \right) dt \right) I(\tau) d\tau,
\end{aligned}
\tag{19}
$$

respectively, with [on the evaluation of c_2 see Appendix B] (in the case $a \neq 0$)

$$
\begin{aligned}
I_0(y) = \check{I}_0(y) &+ \frac{c_2 \sin^2 \kappa y}{2\kappa^2} \\
- 4ac_1 \int_0^y \frac{\sin 2\kappa(y-t)}{2\kappa} &\int_0^t \varepsilon_I(z) dz dt,
\end{aligned}
\tag{20}
$$

$$
ac_1 = -q_s I(0),
\tag{21}
$$

$$
\begin{aligned}
c_2 &= 2I(0)(q_s^2 + q_f^2(0) - p^2) - \frac{2I^2(0)}{1+rI(0)} \\
&+ \frac{2}{r^2} \left(2rI(0) + \frac{1}{1+rI(0)} - \ln(1+rI(0)) \right),
\end{aligned}
\tag{22}
$$

where $\check{I}_0(y)$ is given by equation (15), and with (in the case $a = 0$)

$$
\begin{aligned}
I_0(y) = \check{I}_0(y) &+ \frac{c_2 \sin^2 \kappa y}{2\kappa^2} \\
- 4c_1 \int_0^y \frac{\sin 2\kappa(y-t)}{2\kappa} &\int_0^t \varepsilon_I(z) dz dt,
\end{aligned}
\tag{23}
$$

$$
c_1 = -q_s I(0),
\tag{24}
$$

$$
c_2 = 2I(0)(q_s^2 + q_f^2(0) - p^2),
\tag{25}
$$

where $\check{I}_0(y) = |E_3|^2 \cos(2\kappa y)$.

Iteration of the nonlinear integral equations (16) and (17) leads to a sequence of functions $I_j(y), 0 < y < d$. Subject to certain conditions it can be shown that the limit

$$I(y) = \lim_{j \to \infty} I_j(y)$$

exists uniformly in $0 < y < d$ and represents the unique solution of (16) and (17). The error of approximations can be expressed in terms of the parameters of the problem (see (68) and (69) in Appendix A).

Iterating (16) and (17) once by inserting $I_0(y)$ according to (20) and (23), the first iteration $I_1(y)$ reads ($a \neq 0$)

$$I_1(y) = I_0(y) + \frac{1}{\kappa^2} \int_0^y \sin^2 \kappa(y - \tau) \frac{d\varepsilon_R(\tau)}{d\tau} I_0(\tau) d\tau$$

$$- \frac{2}{\kappa} \int_0^y \sin 2\kappa(y - \tau) \varepsilon_R(\tau) I_0(\tau) d\tau$$

$$- \frac{2}{r\kappa} \int_0^y \sin 2\kappa(y - \tau) I_0(\tau) d\tau$$

$$- \frac{1}{\kappa r^2} \int_0^y \sin 2\kappa(y - \tau) \frac{1}{1 + rI_0(\tau)} d\tau$$

$$+ \frac{1}{\kappa r^2} \int_0^y \sin 2\kappa(y - \tau) \ln(1 + rI_0(\tau)) d\tau$$

$$+ 4 \int_0^y \varepsilon_I(\tau) \left(\int_\tau^y \frac{\sin 2\kappa(y - t)}{2\kappa} \left(\int_\tau^t \varepsilon_I(z) dz \right) dt \right) I_0(\tau) d\tau, \tag{26}$$

and ($a = 0$)

$$I_1(y) = I_0(y) + \frac{1}{\kappa^2} \int_0^y \sin^2 \kappa(y - \tau) \frac{d\varepsilon_R(\tau)}{d\tau} I_0(\tau) d\tau$$

$$- \frac{2}{\kappa} \int_0^y \sin 2\kappa(y - \tau) \varepsilon_R(\tau) I_0(\tau) d\tau$$

$$+ 4 \int_0^y \varepsilon_I(\tau) \left(\int_\tau^y \frac{\sin 2\kappa(y - t)}{2\kappa} \left(\int_\tau^t \varepsilon_I(z) dz \right) dt \right) I_0(\tau) d\tau. \tag{27}$$

$I_1(y)$ is used for numerical evaluation of the physical quantities defined in the following section.

3. Reflectance, transmittance, absorptance, and phase shifts

Conservation of energy requires that absorptance A, transmittance T, and reflectance R are related by

$$A = 1 - R - T, \tag{28}$$

with

$$T = \frac{q_s}{q_c} \frac{I(0)}{a E_0^2}, \quad T = \frac{q_s}{q_c} \frac{I(0)}{E_0^2}, \tag{29}$$

$$R = \frac{|E_r|^2}{E_0^2}, \tag{30}$$

for $a \neq 0$ and $a = 0$, respectively.

Due to the continuity conditions at $y = d$

$$E_0 + |E_r|e^{i\delta_r} = E(d)e^{i\vartheta(d)} \tag{31}$$

$$2E_0 e^{-i\vartheta(d)} = \frac{i}{q_c} \frac{dE(y)}{dy} |_{y=d} + E(d)(1 - \frac{1}{q_c} \frac{d\vartheta(y)}{dy} |_{y=d}) \tag{32}$$

reflectance, transmittance, absorptance and the phase shift on reflection, δ_r, and on transmission, δ_t, can be determined. Combination of equations (31) and (32) yields (for $a \neq 0$)

$$aE_0^2 = \frac{1}{4} \left\{ \frac{\left(\frac{dI(y)}{dy} |_{y=d}\right)^2}{4q_c^2 I(d)} + I(d) \left(1 + \frac{q_s I(0) + \int\limits_0^d \varepsilon_I(\tau)I(\tau)d\tau}{q_c I(d)} \right)^2 \right\}, \tag{33}$$

$$a|E_r|^2 = \frac{1}{4} \left\{ \frac{\left(\frac{dI(y)}{dy} |_{y=d}\right)^2}{4q_c^2 I(d)} + I(d) \left(1 - \frac{q_s I(0) + \int\limits_0^d \varepsilon_I(\tau)I(\tau)d\tau}{q_c I(d)} \right)^2 \right\}, \tag{34}$$

and (for $a = 0$)

$$E_0^2 = \frac{1}{4} \left\{ \frac{\left(\frac{dI(y)}{dy} |_{y=d}\right)^2}{4q_c^2 I(d)} + I(d) \left(1 + \frac{q_s I(0) + \int\limits_0^d \varepsilon_I(\tau)I(\tau)d\tau}{q_c I(d)} \right)^2 \right\}, \tag{35}$$

$$|E_r|^2 = \frac{1}{4} \left\{ \frac{\left(\frac{dI(y)}{dy} |_{y=d}\right)^2}{4q_c^2 I(d)} + I(d) \left(1 - \frac{q_s I(0) + \int\limits_0^d \varepsilon_I(\tau)I(\tau)d\tau}{q_c I(d)} \right)^2 \right\}. \tag{36}$$

Inserting equations (33)-(36) into equation (28) and using equations (29), (33) we obtain

$$A = \frac{1}{q_c a E_0^2} \int\limits_0^d \varepsilon_I(\tau)I(\tau)d\tau, \quad A = \frac{1}{q_c E_0^2} \int\limits_0^d \varepsilon_I(\tau)I(\tau)d\tau, \tag{37}$$

for $a \neq 0$ and $a = 0$, respectively. The continuity conditions (31), (32) and equations (29), (33) and (35) imply

$$\delta_r = -\arcsin \frac{\frac{dI(y)}{dy}\big|_{y=d}}{4q_c a E_0^2 \sqrt{1-T-A}}, \quad \delta_r = -\arcsin \frac{\frac{dI(y)}{dy}\big|_{y=d}}{4q_c E_0^2 \sqrt{1-T-A}} \tag{38}$$

for the phase shift on reflection (for $a \neq 0$ and $a = 0$, respectively), and

$$\delta_t = \vartheta(0) = \int_0^d \frac{q_s I(0) + q_c a E_0^2 \tilde{A}(\tau)}{I(\tau)} d\tau + \arcsin\left(-\frac{\frac{dI(y)}{dy}\big|_{y=d}}{4q_c \sqrt{a E_0^2 I(d)}}\right),$$

$$\delta_t = \vartheta(0) = \int_0^d \frac{q_s I(0) + q_c E_0^2 \tilde{A}(\tau)}{I(\tau)} d\tau + \arcsin\left(-\frac{\frac{dI(y)}{dy}\big|_{y=d}}{4q_c \sqrt{E_0^2 I(d)}}\right), \tag{39}$$

with

$$\tilde{A}(\tau) := \frac{1}{q_c a E_0^2} \int_0^\tau \varepsilon_I(u) I(u) \, du, \quad \tilde{A}(\tau) := \frac{1}{q_c E_0^2} \int_0^\tau \varepsilon_I(u) I(u) \, du, \tag{40}$$

for the phase shift on transmission (for $a \neq 0$ and $a = 0$, respectively).

4. Numerical evaluations

A numerical evaluation of the foregoing quantities is straightforward. It is useful to apply a parametric-plot routine using the first approximation $I_1(y)$. If the parameters of the problem $(a, r, \varepsilon_R, \varepsilon_I, \varepsilon_s, \varepsilon_f^0, \varepsilon_c, p, d, w_0)$ satisfy the convergence conditions (48) and (49) (see Appendix A) the results obtained for $I_1(y)$ are in good agreement with the purely numerical solution of equation (13) and (14) (cf. figure 3).

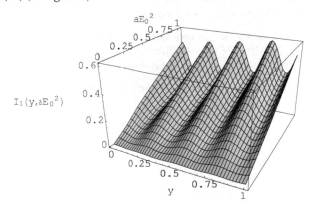

Fig. 2. Dependence of the field intensity $I_1(y, a E_0^2)$ inside the slab on the transverse coordinate y and $a E_0^2$ for
$r = 1000$, $\varepsilon_I = 0.1$, $\varepsilon_c = 1$, $\varepsilon_s = 1.7$, $\varepsilon_f^0 = 3.5$, $\varphi = 1.107$, $d = 1$, $\gamma = 0.033$, $b = 0.1$.

For the numerical evaluations the following steps can be performed:

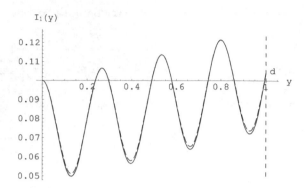

Fig. 3. Dependence of the field intensity $I_1(y, aE_0^2)$ inside the slab on the transverse coordinate y for $a|E_3|^2 = 0.1$. The other parameters are as in figure 2. Solid curve corresponds to the first iteration of equation (26) and dashed curve to the numerical solution of the system of differential equations (6), (7).

(i) Prescribe the parameters of the problem such that (48) and (49) are satisfied.

(ii) Prescribe a certain upper bound (accuracy) of the right-hand side R_j (see (66) in Appendix A) and perform a parametric plot of R_j (with $I(0)$ as parameter) with j=1. If R_1 is smaller (or equal) than (to) the prescribed accuracy for all aE_0^2 (or E_0^2) of a certain interval, accept $I_1(y)$ as a suitable approximation.

(iii) If R_1 exceeds the prescribed accuracy calculate $I_2(y)$ according to (50) and check again according to step (ii) or enlarge the accuracy so that R_1 is smaller (or equal) than (to) the prescribed accuracy.

The reason for the satisfactory agreement between the exact numerical solution and the first approximation $I_1(y)$ (cf. figure 3) is due to the foregoing explanation.

If $a|E_3|^2$ (or $|E_3|^2$) is fixed (as in the numerical example below), and thus aE_0^2 (or E_0^2) according to (33) (or (35)), the inequality (68) (or (69)) can be used to optimize the iteration approach with respect to another free parameter, e.g., d or r or p, as indicated.

Using the first approximation the phase function can be evaluated according to equation (8) as (for all values of a)

$$\vartheta_1(y) = \vartheta_1(d) - q_s I(0) \int_d^y \frac{d\tau}{I_1(\tau)} - \int_d^y \frac{d\tau}{I_1(\tau)} \int_0^\tau \varepsilon_I(\xi) I_1(\xi) d\xi, \tag{41}$$

where

$$\sin \vartheta_1(d) = - \frac{\frac{dI_1(y)}{dy}|_{y=d}}{4q_c \sqrt{aE_0^2 I_1(d)}},$$

$$\sin \vartheta_1(d) = - \frac{\frac{dI_1(y)}{dy}|_{y=d}}{4q_c \sqrt{E_0^2 I_1(d)}}$$

$$\tag{42}$$

for $a \neq 0$ and $a = 0$, respectively.

Thus, the approximate solution of the problem is represented by equations (26) or (27) and (41). The appropriate parameter is $I(0) = aE^2(0)$ or $I(0) = E^2(0)$, since E_0 in equation (42) can be expressed in terms of $I(0)$ as shown in (33) or (35).

For illustration we assume a permittivity according to ($a \neq 0$)

$$\varepsilon_f(y) = \varepsilon_f^0 + \varepsilon_R(y) + i\varepsilon_I + \frac{I(y)}{1 + rI(y)}, \tag{43}$$

with

$$\varepsilon_R(y) = \gamma \cos^2 \frac{by}{d}, \tag{44}$$

where $\varepsilon_f^0, \gamma, b, d, r$ are real constants. For simplicity, ε_I is also assumed to be constant. Results for the first iterate solution $I_1(y, aE_0^2)$ are depicted in figures 2, 3. Using $I_1(y, aE_0^2)$, the phase function $\vartheta_1(y, aE_0^2)$, absorptance $A_1(d, aE_0^2)$ and phase shift on reflection $\delta_{r1}(y, aE_0^2)$ are shown in figures 4, 5 and 6, respectively. The left hand side of condition (48) is 0.572 for the parameters selected in this example. Results for R, T and the phase shift on transmission can be obtained similarly.

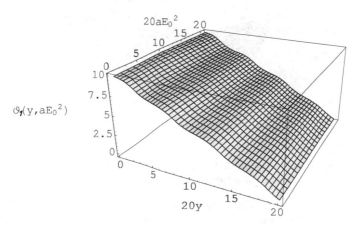

Fig. 4. Phase function $\vartheta_1(y, aE_0^2)$ according to equation (41) inside the slab. Parameters as in figure 3.

5. Summary

Based on known mathematics we have proposed an iterative approach to the scattering of a plane TE-polarized optical wave at a dielectric film with permittivities modeled by a complex continuously differentiable function of the transverse coordinate.

The result is an approximate analytical expression for the field intensity inside the film that can be used to express the physical relevant quantities (reflectivity, transmissivity, absorptance, and phase shifts). Comparison with exact numerical solutions shows satisfactory agreement.

It seems appropriate to explain the benefits of the present approach:

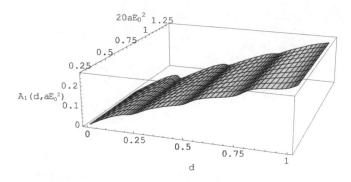

Fig. 5. Absorptance A_1 depending on the layer thickness d and on the incident field intensity aE_0^2 for the same parameters as in figure 3.

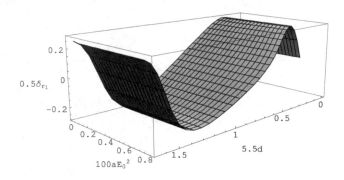

Fig. 6. Phase shift on reflection δ_{r1} depending on the layer thickness d and on the incident field intensity aE_0^2 for the same parameters as in figure 3.

(i) The approach yields (approximate) solutions in cases where the usual methods (cf. Refs. (Chen & Mills, 1987; 1988; Leung, 1985; 1988; Peschel, 1988; Schürmann & Schmoldt, 1993)) fail or could not be applied till now.

(ii) The quality of the approximate solutions can be estimated in dependence on the parameters of the problem.

On the other hand the conditions of convergence explicitly depend on the permittivity functions in question and thus have to be derived for every permittivity anew (cf. (Serov et al., 2004; 2010) and (65)).

6. Acknowledgment

Financial support by the Deutsche Forschungsgemeinschaft (Graduate College 695 "Nonlinearities of optical materials") is gratefully acknowledged. One of the authors (V.S.S.) gratefully acknowledges the support by the Academy of Finland (Application No. 213476, Finnish Programme for Centres of Excellence in Research 2006-2011).

7. Appendix

7.1 Appendix A

We introduce in the Banach space $C[0, d]$ bounded integral operators $N_1, N_2, N_3, N_4, N_5, N_6$ by

$$N_1(I) = \frac{1}{\kappa^2} \int_0^y \sin^2 \kappa(y - \tau) \frac{d\varepsilon_R(\tau)}{d\tau} I(\tau) d\tau,$$

$$N_2(I) = -\frac{2}{\kappa} \int_0^y \sin 2\kappa(y - \tau)\varepsilon_R(\tau) I(\tau) d\tau,$$

$$N_3(I) = -\frac{2}{\kappa} \int_0^y \sin 2\kappa(y - \tau) I(\tau) d\tau,$$

$$N_4(I) = -\frac{1}{\kappa} \int_0^y \sin 2\kappa(y - \tau) \frac{1}{1 + rI(\tau)} d\tau,$$

$$N_5(I) = \frac{1}{\kappa} \int_0^y \sin 2\kappa(y - \tau) \ln(1 + rI(\tau)) d\tau,$$

$$N_6(I) = 4 \int_0^y \varepsilon_I(s)\psi(y, s) I(s) ds, \tag{45}$$

where

$$\psi(y, s) = \int_s^y \frac{\sin 2\kappa(y - t)}{2\kappa} \left(\int_s^t \varepsilon_I(\tau) d\tau \right) dt, \tag{46}$$

with the values $\|N_1\|, \|N_2\|, \|N_3\|, \|N_4\|, \|N_5\|, \|N_6\|, \|I_0\|$ which are defined as

$$\|N_1\| = \frac{1}{\kappa^2} \max_{0 \le y \le d} \int_0^y |\sin^2 \kappa(y - \tau)| \cdot \left|\frac{d\varepsilon_R(\tau)}{d\tau}\right| d\tau,$$

$$\|N_2\| = \frac{2}{\kappa} \max_{0 \le y \le d} \int_0^y |\sin 2\kappa(y - \tau)| \cdot |\varepsilon_R(\tau)| d\tau,$$

$$\|N_3\| = \frac{2}{\kappa} \max_{0 \le y \le d} \int_0^y |\sin 2\kappa(y - \tau)| d\tau,$$

$$\|N_4\| = \frac{1}{\kappa} \max_{0 \le y \le d} \int_0^y |\sin 2\kappa(y - \tau)| d\tau,$$

$$\|N_5\| = \frac{1}{\kappa} \max_{0 \le y \le d} \int_0^y |\sin 2\kappa(y - \tau)| d\tau = \|N_4\|,$$

$$\|N_6\| = 4 \max_{0 \le y \le d} \int_0^y |\varepsilon_I(z)| \cdot |\psi(y, z)| dz,$$

$$\|I_0\| = \max_{0 \le y \le d} |I_0|. \tag{47}$$

We are in the position now to show that if

$$\|N_1\| + \|N_2\| + \|N_6\| + \frac{\|N_3\| + 2\|N_4\|}{r} < 1 \tag{48}$$

and

$$\frac{\|I_0\| + \frac{1}{r^2}\|N_4\|}{1 - (\|N_1\| + \|N_2\| + \|N_6\| + \frac{\|N_3\| + \|N_4\|}{r})} < \rho, \tag{49}$$

then in any ball $S_\rho(0)$ there exists a unique solution of the nonlinear integral equation (18) and this solution can be obtained as a uniform limit

$$I(y) = \lim_{j \to \infty} I_j(y)$$

of the iterations of (18). Indeed, let us introduce the iterations of (18) as follows:

$$
\begin{aligned}
I_j(y) = I_0(y) &+ \frac{1}{\kappa^2} \int_0^y \sin^2 \kappa(y - \tau) \frac{d\varepsilon_R(\tau)}{d\tau} I_{j-1}(\tau) d\tau \\
&- \frac{2}{\kappa} \int_0^y \sin 2\kappa(y - \tau) \varepsilon_R(\tau) I_{j-1}(\tau) d\tau \\
&- \frac{2}{r\kappa} \int_0^y \sin 2\kappa(y - \tau) I_{j-1}(\tau) d\tau \\
&- \frac{1}{\kappa r^2} \int_0^y \sin 2\kappa(y - \tau) \frac{1}{1 + r I_{j-1}(\tau)} d\tau \\
&+ \frac{1}{\kappa r^2} \int_0^y \sin 2\kappa(y - \tau) \ln(1 + r I_{j-1}(\tau)) d\tau \\
&+ 4 \int_0^y \varepsilon_I(\tau) \left(\int_\tau^y \frac{\sin 2\kappa(y - t)}{2\kappa} \left(\int_\tau^t \varepsilon_I(z) dz \right) dt \right) I_{j-1}(\tau) d\tau,
\end{aligned}
\tag{50}
$$

where $j = 1, 2, \ldots$, and $I_0(y)$ is given by equation (20). In order to prove that the sequence (50) is uniformly convergent to the solution of (16) it suffices to check that all conditions of the Banach Fixed-Point Theorem (see (Zeidler, 1995)) are fulfilled.

We consider the nonlinear operator F as

$$F(I) := I_0(y) + N_1(I) + N_2(I) + \frac{1}{r} N_3(I) + \frac{1}{r^2} N_4(I) + \frac{1}{r^2} N_5(I) + N_6(I). \tag{51}$$

Then equation (18) can be rewritten in operator form

$$I(y) = F(I)(y). \tag{52}$$

We consider ρ such that $\|I\| = \max_{0 \le y \le d} I(y) \le \rho$. First we must check whether this operator F maps the ball $S_\rho(0)$ to itself. Indeed, if $I(y) \in S_\rho(0)$ then

$$
\begin{aligned}
\|F(I)\| \le \|I_0\| &+ \|N_1\| \cdot \|I\| + \|N_2\| \cdot \|I\| + \|N_6\| \cdot \|I\| + \frac{1}{r} \|N_3\| \cdot \|I\| \\
&+ \frac{1}{r^2} \|N_4\| \cdot \frac{1}{1 + r \min_{0 \le y \le d} I(y)} + \frac{1}{r^2} \|N_4\| \cdot r \|I\| \\
\le \|I_0\| &+ \|N_1\| \cdot \rho + \|N_2\| \cdot \rho + \|N_6\| \cdot \rho + \frac{1}{r} \|N_3\| \cdot \rho + \frac{1}{r^2} \|N_4\| \\
&+ \frac{1}{r} \|N_4\| \cdot \rho.
\end{aligned}
\tag{53}
$$

Thus, the following inequality must be valid

$$\|I_0\| + \frac{1}{r^2} \|N_4\| + \left(\|N_1\| + \|N_2\| + \|N_6\| + \frac{1}{r} (\|N_3\| + \|N_4\|) \right) \cdot \rho < \rho. \tag{54}$$

This inequality holds if

$$\frac{\|I_0\| + \frac{1}{r^2}\|N_4\|}{1 - (\|N_1\| + \|N_2\| + \|N_6\| + \frac{\|N_3\| + \|N_4\|}{r})} < \rho, \tag{55}$$

and thus if

$$\|N_1\| + \|N_2\| + \|N_6\| + \frac{\|N_3\| + \|N_4\|}{r} < 1. \tag{56}$$

It means that for this value of ρ continuous map F transfers ball $S_\rho(0)$ in itself. Hence, equation (16) has at least one solution inside $S_\rho(0)$. For uniqueness of this solution it remains to prove that F is contractive (see (Zeidler, 1995)). To prove the contraction of F we consider

$$F(I_1) - F(I_2) = N_1(I_1 - I_2) + N_2(I_1 - I_2) + N_6(I_1 - I_2)$$
$$+ \frac{1}{r}N_3(I_1 - I_2) + \frac{1}{r^2}(N_4(I_1) - N_4(I_2)) + \frac{1}{r^2}(N_5(I_1) - N_5(I_2)). \tag{57}$$

Hence

$$\|F(I_1) - F(I_2)\| \le \|N_1\|\|I_1 - I_2\| + \|N_2\|\|I_1 - I_2\| + \|N_6\|\|I_1 - I_2\|$$
$$+ \|N_3\|\|\frac{1}{r}(I_1 - I_2)\| + \|\frac{1}{r^2}(N_4(I_1) - N_4(I_2))\|$$
$$+ \|\frac{1}{r^2}(N_5(I_1) - N_5(I_2))\|. \tag{58}$$

The following estimations hold

(i) $\quad \|\frac{1}{r^2}(N_4(I_1) - N_4(I_2))\| \le \max\limits_{0 \le y \le d} \frac{1}{\kappa r^2} \int_0^y |\sin 2\kappa(y - \tau)| \cdot$

$\left| \frac{1}{1 + rI_1(\tau)} - \frac{1}{1 + rI_2(\tau)} \right| d\tau =$

$\max\limits_{0 \le y \le d} \frac{1}{\kappa r^2} \int_0^y |\sin 2\kappa(y - \tau)| \cdot \left| \frac{rI_2(\tau) - rI_1(\tau)}{(1 + rI_1(\tau))(1 + rI_2(\tau))} \right| d\tau$

$\le \frac{1}{r} \max\limits_{0 \le y \le d} \frac{1}{\kappa} \int_0^y |\sin 2\kappa(y - \tau)| d\tau \cdot \|I_1 - I_2\|,$

hence

$$\|\frac{1}{r^2}(N_4(I_1) - N_4(I_2))\| \le \frac{\|N_4\|}{r} \cdot \|I_1 - I_2\|. \tag{59}$$

(ii) $\quad \|\frac{1}{r^2}(N_5(I_1) - N_5(I_2))\| \le \max\limits_{0 \le y \le d} \frac{1}{\kappa r^2} \int_0^y |\sin 2\kappa(y - \tau)| \cdot$

$|\ln(1 + rI_1(\tau)) - \ln(1 + rI_2(\tau))| d\tau. \tag{60}$

Using

$$|\ln(1 + rI_1) - \ln(1 + rI_2)| = \left| \ln\frac{1 + rI_1}{1 + rI_2} \right|$$
$$= \left| \ln\left(1 + \frac{r(I_1 - I_2)}{1 + rI_2}\right) \right| \le r\|I_1 - I_2\|, \tag{61}$$

equation (60) yields

$$\left\|\frac{1}{r^2}(N_5(I_1) - N_5(I_2))\right\| \leq \frac{1}{r^2} \cdot \|N_4\| \cdot r \cdot \|I_1 - I_2\| = \frac{\|N_4\|}{r}\|I_1 - I_2\|. \tag{62}$$

Thus, from equation (58), one obtains

$$\|F(I_1) - F(I_2)\| \leq (\|N_1\| + \|N_2\| + \|N_6\| + \frac{\|N_3\|}{r} + \frac{\|N_4\|}{r} + \frac{\|N_4\|}{r})$$
$$\cdot \|I_1 - I_2\|$$
$$= (\|N_1\| + \|N_2\| + \frac{\|N_3\| + 2\|N_4\|}{r}) \cdot \|I_1 - I_2\|, \tag{63}$$

so that F is contractive if

$$\|N_1\| + \|N_2\| + \|N_6\| + \frac{\|N_3\| + 2\|N_4\|}{r} < 1. \tag{64}$$

Thus, the uniform convergence follows.

If we denote by m the left-hand side of the inequality (48) the solution $I(y)$ of (16) can be approximated by the iterations $I_j(y)$ as follows (see (Zeidler, 1995)):

$$\|I - I_j\| \leq \frac{m^j}{1 - m}\|I_1 - I_0\|$$
$$\leq \frac{m^j}{1 - m}\left(\frac{1}{\kappa r^2}\max_{0 \leq y \leq d}\int_0^y |\sin 2\kappa(y - \tau)|d\tau + m\|I_0\|\right)$$
$$\leq \frac{m^j}{1 - m}\left(\frac{d^2}{r^2} + m\|I_0\|\right), \tag{65}$$

where $j = 0, 1, 2, \ldots$ and I_0 is defined in (20).

Let us remark that for the sufficient condition (48) to hold parameters must be chosen such that (48) holds even if r is small (Equation (18) represents the exact solution $I(y)$ if (48) and (49) are satisfied). $I(y)$ can be approximated by the first iteration $I_1(y)$ with the error $\frac{d^2}{r^2}\frac{m}{1-m} + \frac{m^2}{1-m}\|I_0\|$, where m denotes the left-hand side of (48). Condition (48) must hold for a particular $r > 0$. In the limit $r \to 0$ equation (18) transforms to equation

$$I(y) = I_0(y) + \frac{1}{\kappa^2}\int_0^y \sin^2 \kappa(y - \tau)\frac{d\varepsilon_R(\tau)}{d\tau}I(\tau)d\tau$$
$$- \frac{2}{\kappa}\int_0^y \sin 2\kappa(y - \tau)\varepsilon_R(\tau)I(\tau)d\tau - \frac{3}{2\kappa}\int_0^y \sin 2\kappa(y - \tau)I^2(\tau)d\tau$$
$$+ 4\int_0^y \varepsilon_I(\tau)\left(\int_\tau^y \frac{\sin 2\kappa(y - t)}{2\kappa}\left(\int_\tau^t \varepsilon_I(z)dz\right)dt\right)I(\tau)d\tau, \tag{66}$$

where $I_0(y)$ is the same as (23) with the constant c_2 which is equal to

$$c_2 = I_0(0)(q_s^2 + q_f^2(0) - p^2) - 2I^2(0).$$

Equation (66) is equivalent to (in the case of lossless medium) (41) in (Serov et al., 2004). Equation (66) is uniquely solvable in the ball of radius ρ if the following conditions are satisfied (they are consistent with the corresponding conditions from (Serov et al., 2004; 2010)):

$$m + 3d^2\rho < 1, \quad m + d\sqrt{6\|I_0\|} < 1,$$

where $m = \|N_1\| + \|N_2\| + \|N_c\|$ and the radius ρ is chosen so that

$$\rho \geq \frac{1 - m - \sqrt{(1-m)^2 - 6d^2\|I_0\|}}{3d^2}.$$

In order to obtain a condition of the type (48) for all $0 < r < 1$ (uniformly) combination of N_3, N_4, N_5 and part of I_0 within the estimations is necessary. It seems impossible to obtain a condition of the type (48) uniformly with respect to all nonnegative r. It is possible only to obtain such kind of condition uniformly for $0 < r < 1$ or for $1 < r < \infty$ independently. In this respect, some mathematical complications arise that are not the main point of this paper.

Estimation of $\|I_0\|$ (cf. Appendix C) gives

$$\|I_0\| \leq I(0) + \frac{1}{2}|c_2|d^2 + \frac{2}{3}a|c_1|\|\varepsilon_I\|d^3,$$

$$\|I_0\| \leq I(0) + \frac{1}{2}|c_2|d^2 + \frac{2}{3}|c_1|\|\varepsilon_I\|d^3, \tag{67}$$

where constants c_1 and c_2 are defined by (21) and (22) for $a \neq 0$, and by (24) and (25) for $a = 0$, respectively. Combining (65) and (67) we obtain the error of approximation (for $a \neq 0$)

$$R_j := \|I - I_j\| \leq \frac{d^2}{r^2} \cdot \frac{m^j}{1-m} + \frac{m^{j+1}}{1-m}\left(I(0) + \frac{1}{2}|c_2|d^2 + \frac{2}{3}a|c_1|\|\varepsilon_I\|d^3\right), \tag{68}$$

where $j = 0, 1, 2, \ldots$.

Since for linear case ($a = 0$) equation (17) is the linear Volterra integral equation this equation has always a unique solution and the following error of approximation holds:

$$\|I - I_j\| \leq \frac{(\|I_0\|md)^{j+1}}{(j+1)!}e^{\|I_0\|md}, \tag{69}$$

where $j = 0, 1, 2, \ldots$, $\|I_0\|$ is estimated in (67), $m = \|N_1\| + \|N_2\| + \|N_6\|$ and I_j are defined by

$$I_j(y) = I_0(y) + \frac{1}{\kappa^2}\int_0^y \sin^2\kappa(y-\tau)\frac{d\varepsilon_R(\tau)}{d\tau}I_{j-1}(\tau)d\tau$$

$$-\frac{2}{\kappa}\int_0^y \sin 2\kappa(y-\tau)\varepsilon_R(\tau)I_{j-1}(\tau)d\tau$$

$$+4\int_0^y \varepsilon_I(\tau)\left(\int_\tau^y \frac{\sin 2\kappa(y-t)}{2\kappa}\left(\int_\tau^t \varepsilon_I(z)dz\right)dt\right)I_{j-1}(\tau)d\tau. \tag{70}$$

7.2 Appendix B

The constant of integration c_2 is determined by equations (13) and (14) for $a \neq 0$ and for $a = 0$, respectively with $y = 0$ as

$$c_2 = \left.\frac{d^2 I(y)}{dy^2}\right|_{y=0} + 4(q_f^2(0) - p^2)I(0)$$

$$+ \frac{2}{r^2}\left(2rI(0) + \frac{1}{1+rI(0)} - \ln(1+rI(0))\right), \tag{71}$$

$$c_2 = \left.\frac{d^2 I(y)}{dy^2}\right|_{y=0} + 4(q_f^2(0) - p^2)I(0). \tag{72}$$

According to equation (10), the second derivative of the field intensity $I(y)$ at $y = 0$ is given by (for $a \neq 0$)

$$\left.\frac{d^2 I(y)}{dy^2}\right|_{y=0} = 2q_s^2 I(0) - 2(q_f^2(0) - p^2)I(0) - \frac{2I^2(0)}{1+rI(0)}, \tag{73}$$

and (for $a = 0$)

$$\left.\frac{d^2 I(y)}{dy^2}\right|_{y=0} = 2q_s^2 I(0) - 2(q_f^2(0) - p^2)I(0), \tag{74}$$

leading to, taking into account boundary conditions, $E(0) = E_3 e^{-i\vartheta(0)}$ and $\left.\frac{dE(y)}{dy}\right|_{y=0} = 0$,

$$c_2 = 2q_s^2 I(0) + 2(q_f^2(0) - p^2)I(0) - \frac{2I^2(0)}{1+rI(0)}$$

$$+ \frac{2}{r^2}\left(2rI(0) + \frac{1}{1+rI(0)} - \ln(1+rI(0))\right), \tag{75}$$

$$c_2 = 2q_s^2 I(0) + 2(q_f^2(0) - p^2)I(0). \tag{76}$$

for $a \neq 0$ and $a = 0$, respectively.

7.3 Appendix C

With $\varepsilon_R(x) \in C^1[0,d]$ and $\varepsilon_I(x) \in C[0,d]$ one obtains

$$\|N_1\| = \frac{1}{\kappa^2} \max_{0 \leq y \leq d} \int_0^y |\sin^2 \kappa(y-\tau)| \cdot |\varepsilon_R'(\tau)| d\tau$$

$$\leq \max_{0 \leq y \leq d} \int_0^y (y-\tau)^2 d\tau \cdot \|\varepsilon_R'\| = \frac{1}{3}d^3 \|\varepsilon_R'\|, \tag{77}$$

$$\|N_2\| = \frac{2}{\kappa} \max_{0 \leq y \leq d} \int_0^y |\sin 2\kappa(y-\tau)| \cdot |\varepsilon_R(\tau)| d\tau$$

$$\leq 4 \max_{0 \leq y \leq d} \int_0^y (y-\tau) d\tau \cdot \|\varepsilon_R\| = 2d^2 \|\varepsilon_R\|, \tag{78}$$

$$\|N_3\| = \frac{2}{\kappa} \max_{0 \leq y \leq d} \int_0^y |\sin 2\kappa(y-\tau)| d\tau \leq 4 \max_{0 \leq y \leq d} \int_0^y (y-\tau) d\tau = 2d^2, \tag{79}$$

$$\|N_4\| = \|N_5\| = \frac{1}{\kappa} \max_{0 \leq y \leq d} \int_0^y |\sin 2\kappa(y-\tau)| d\tau$$

$$\leq 2 \max_{0 \leq y \leq d} \int_0^y (y-\tau) d\tau = d^2, \tag{80}$$

$$\|N_6\| = 4 \max_{0 \leq y \leq d} \int_0^y |\varepsilon_I(z)| \cdot |\psi(y,z)| dz$$

$$\leq 4\|\varepsilon_I\| \max_{0 \leq y \leq d} \int_0^y \int_z^y \frac{|\sin 2\kappa(y-\tau)|}{2\kappa} \int_z^t |\varepsilon_I(\tau)| d\tau dt dz$$

$$\leq 4\|\varepsilon_I\|^2 \max_{0 \leq y \leq d} \int_0^y \int_z^y (y-t)(t-z) dt dz = \frac{d^4}{6} \|\varepsilon_I\|^2, \tag{81}$$

$$\|I_0\| \leq \max_{0 \leq y \leq d} a|E_3|^2 \cdot |\cos 2\kappa y| + |c_2| \max_{0 \leq y \leq d} \frac{|\sin^2 \kappa y|}{2\kappa^2}$$

$$+ 4a|c_1| \max_{0 \leq y \leq d} \int_0^y \frac{|\sin 2\kappa(y-t)|}{2\kappa} \int_0^t |\varepsilon_I(\tau)| d\tau dt$$

$$\leq a|E_3|^2 + \frac{1}{2}|c_2|d^2 + 4a|c_1| \cdot \|\varepsilon_I\| \max_{0 \leq y \leq d} \int_0^y (y-t) t dt$$

$$= a|E_3|^2 + \frac{1}{2}|c_2|d^2 + \frac{2}{3}a|c_1| \cdot \|\varepsilon_I\| d^3, \tag{82}$$

where $a \neq 0$ and ε'_R denotes the first derivative of ε_R.

For $a = 0$ the estimate of $\|I_0\|$ (82) transforms to the following one

$$\|I_0\| \leq |E_3|^2 + \frac{1}{2}|c_2|d^2 + \frac{2}{3}|c_1| \cdot \|\varepsilon_I\| d^3, \tag{83}$$

with c_1 and c_2 from (24) and (25).

For ε_R, given by (44), we obtain

$$\|\varepsilon_R\| \leq \gamma, \qquad \|\varepsilon'_R\| \leq \begin{cases} \frac{2\gamma b^2}{d}, & 2b \leq 1, \\ \frac{\gamma b}{d}, & 2b > 1. \end{cases} \tag{84}$$

8. References

Bang, O.; Krolikowski, W.; Wyller, J. & Rasmussen, J.J. (2002). Collapse arrest and soliton stabilization in nonlocal nonlinear media. *Physical Review E*, Vol. 66, 046619

Berge, L.; Gouedard, C.; Schjodt-Eriksen, J. & Ward, H. (2003). Filamentation patterns in Kerr media vs. beam shape robustness, nonlinear saturation and polarization state. *Physica D*, Vol. 176, 181–211

Chen, W. & Mills, D. L. (1987). Optical response of a nonlinear dielectric film. *Phys. Rev. B*, Vol. 35, 524–532

Chen, W. & Mills, D. L. (1988). Optical behavior of a nonlinear thin film with oblicue S-polarized incident wave. *Phys. Rev. B*, Vol. 38, 12814–12822

Dreischuh, A.; Paulus, G.G.; Zacher, F.; Grasbon, F.; Neshev, D. & Walther, H. (1999). Modulational instability of multi-charged optical vortex solitons under saturation on the nonlinearity. *Physical Review E*, Vol. 60, 7518–7522

Gordillo-Vázquez, F. J. & Pecharromán, C. (2003). An effective-medium approach to the optical properties of heterogeneous materials with nonlinear properties. *J. Mod. Optics*, Vol. 50(1), 113–135

Kartashov, Y.V.; Vysloukh, V.A. & Torner, L. (2003). Two-dimensional cnoidal waves in Kerr-type saturable nonlinear media. *Physical Review E*, Vol. 68, 015603(R)

Leung, K. M. (1985). P-polarized nonlinear surface polaritons in materials with intensity-dependent dielectric functions. *Phys. Rev. B*, Vol. 32(8), 5093–5101

Leung, K. M. (1988). Scattering of transverse-electric electromagnetic waves with a finite nonlinear film. *J. Opt. Soc. Am. B*, Vol. 5(2), 571–574

Leung, K. M. (1989). Exact results for the scattering of electromagnetic waves with a nonlinear film,. *Phys. Rev. B*, Vol. 39, 3590–3598

Peschel, Th. (1988). Investigation of optical tunneling through nonlinear films. *J. Opt. Soc. Am. B*, Vol. 5(1), 29–36

Schürmann, H. W. & Schmoldt, R. (1993). On the theory of reflectivity and transmissivity of a lossless nonlinear dilectric slab. *Z. Phys. B*, Vol. 92, 179–186

Schürmann, H. W. & Schmoldt, R. (1996). Optical response of a nonlinear absorbing dielectric film. *Optics Letters*, Vol. 21 (6), 387–389

Schürmann, H. W.; Serov, V. S. & Shestopalov, Yu. V. (2001). Reflection and transmission of a plane TE-wave at a lossless nonlinear dielectric film. *Physica D*, Vol. 158, 197–215

Serov, V. S.; Schürmann, H. W. & Svetogorova, E. (2004). Integral equation approach to reflection and transmission of a plane TE-wave at a (linear/nonlinear) dielectric film with spatially varying permittivity. *J. of Phys. A: Math. Gen.*, Vol. 37 (10), 3489–3500

Serov, V. S.; Schürmann, H. W. & Svetogorova, E. (2010). Application of the Banach fixed-point theorem to the scattering problem at a nonlinear three-layer structure with absorption. *Fixed Point Theory and Appl.*, Vol. 2010, 493682

Stakgold, I. (1967). *Boundary Value Problems in Mathematical Physics vol 1*, Macmillan, New York

Yuen, K.P. & Yu, K. W. (1997). Optical response of a nonlinear composite film. *J. Opt. Soc. Am. B*, Vol. 14, 1387–1389

Zeidler, E. (1995). *Applied functional analysis [Part I]: Applications to mathematical physics*, Springer-Verlag, New York

Part 3

Nonlinear Optical Materials, Different New Approach and Ideas

7

Epitaxial (Ba,Sr)TiO$_3$ Ferroelectric Thin Films for Integrated Optics

D. Y. Wang and S. Li

School of Materials Science and Engineering,
The University of New South Wales, Sydney,
Australia

1. Introduction

Complex ferroelectric oxides with excellent optical properties and strong electro-optic (E-O) effects have now opened up the potential for guided-wave devices used in multifunctional integrated optics (Wessels, 2007). Since an applied electric field can result in a change in both the dimensions and orientation of the index ellipsoid of the material through the E-O effect, the E-O effect affords convenient and widely used means of controlling the phase or intensity of the optical radiation. Electro-optic modulators, in particular operating at near infrared wavelengths of 1-1.6 μm, are essential for high-speed and wide bandwidth optical communication systems and ultrafast information processing applications (Tang et al., 2005). Bulk ferroelectric single crystals and transparent ceramics, e.g. LiNbO$_3$ (Gopalakrishnan et al., 1994; Wooten et al., 2000) and (Pb,La)(Zr,Ti)O$_3$ (Chen et al., 1980; Haertling, 1987), are commonly used. However, the realization of thin film E-O devices is of strong scientific and technological interest, since they require relatively low driving power and possess higher interaction efficiency in comparison to bulk modulators. The use of thin films in E-O devices can clearly lead to geometrical flexibility and ability to grow waveguides on diverse substrates for possible integration with existing semiconductor technologies to produce devices much smaller than bulk hybrid counterparts.

In spite of its promise, ferroelectric thin films for optical applications are not well developed. This results from numerous requirements on material perfection for optical devices. At present, only a few laboratories have succeeded to partially master the thin film technology for the optical devices (Guarino et al. 2007; Masuda et al. 2011; Nakada et al. 2009; Petraru et al. 2002; Suzuki et al. 2008), but steady progress is visible. A typical thin film E-O modulator concept comprises: (1) a ferroelectric thin film layer with high E-O coefficient; (2) for waveguide devices, the ferroelectric thin film needs to be optically transparent preferably with low optical loss. Thus heteroepitaxial deposition is required; (3) the epitaxial ferroelectric film must be deposited on substrates with a lower refractive index than the layer itself for a strong light confinement and high optical power density; (4) the waveguide modulator structure, for instance, a Mach-Zehnder geometry, has to be patterned by lithography and etching and also top cladding and electrodes need to be deposited.

Great efforts have been made to explore suitable ferroelectric oxide materials that have large E-O coefficient and can be epitaxially grown on low-refractive-index substrates. To date, the most widely studied ferroelectrics for E-O applications are the titanates and niobates, including $BaTiO_3$ (Kim & Kwok, 1995), $PbZr_xTi_{1-x}O_3$ (Kang et al., 2008; Zhu et al., 2010), (Pb,La)(Zr,Ti)O_3 (Adachi & Wasa, 1991; Masuda et al., 2010; Uchiyama et al., 2007), $LiNbO_3$ (Lee et al., 1996), $KNbO_3$ (Graettinger et al. 1991), (K,Na)NbO_3 (Blomqvist et al., 2005), $KTa_xNb_{1-x}O_3$ (Hoerman et al., 2003), $Sr_xBa_{1-x}Nb_2O_6$ (Tayebati et al., 1996) and $Pb(Mg_{1/3}Nb_{2/3})O_3$-$PbTiO_3$ (Lu et al., 1998, 1999) etc. Recently, the deposition of ferroelectric oxide thin films by a variety of techniques have been explored including molecular beam epitaxy (MBE), pulsed laser deposition (PLD), sputtering, sol-gel and metal-organic chemical vapor deposition (MOCVD) (Wessels, 2007). In contrast to other methods, PLD permits a stoichiometric transfer of material from the target to the film and film growth at high temperatures in reactive ambient gas, in particular, oxygen. However, to produce high quality thin films with good optical transparency and low optical loss for waveguide applications is still a very challenging task. $Ba_{1-x}Sr_xTiO_3$, abbreviated as BST, traditionally considered as an excellent microwave material for wide applications in wireless communication due to its large dielectric tunability at GHz regime, attracts much attention in optoelectronic community because of its high E-O coefficients (Kim et al., 2003; Li et al., 2000; Wang et al., 2007a, 2010). For integration, the films are not only required to have E-O properties that are comparable to those of the bulk but also must have a high degree of microstructure perfection in order to minimize optical scattering losses. In this case, the optical propagation loss has been the most serious barrier for practical applications of ferroelectric thin films to waveguide devices. Hence it must be reduced below a level of about 1 dB/cm for practical applications (Wessels et al., 1996). $Ba_{1-x}Sr_xTiO_3$ thin films have been proved to possess high optical transparency and acceptable optical loss (Wang et al., 2006a), which makes $Ba_{1-x}Sr_xTiO_3$ a very promising candidate for active waveguide applications. Furthermore, $Ba_{1-x}Sr_xTiO_3$ thin films have potential to overcome the major drawbacks of E-O ferroelectric materials, such as the high cost and long optical path length of $LiNbO_3$ and $LiTaO_3$ single crystals and the environmental burden of lead content in (Pb,La)(Zr,Ti)O_3 transparent ceramics and thin films.

In this chapter, $Ba_{0.7}Sr_{0.3}TiO_3$ ferroelectric thin films with large E-O effect were epitaxially grown on single-crystal MgO substrates by pulsed laser deposition technique and the important issues of their material properties are tackled. $Ba_{0.7}Sr_{0.3}TiO_3$/MgO rib-type waveguides and Mach-Zehnder modulators are designed, fabricated and characterized.

2. $Ba_{0.7}Sr_{0.3}TiO_3$ thin film deposition

Orientation engineered $Ba_{0.7}Sr_{0.3}TiO_3$ thin films with surface normal orientations of [001], [011] and [111] have been epitaxially deposited on optically double-side polished single-crystal MgO [001], [011] and [111] substrates, respectively, by pulsed laser deposition using a KrF excimer laser (Lambda Physik COMPex 205) with a wavelength of 248 nm, a pulse width of 28 ns and a repetition rate of 10 Hz. The laser beam impacts the rotating stoichiometric target with an energy density of 2 J /cm². The distance between the target and the substrate was fixed at 5 cm, while the substrate temperature was maintained at 750 °C. All films were prepared in an oxygen atmosphere with partial pressure of 27 Pa. The growth conditions used in this work have been optimized and summarized in Table 1.

The crystal structures of the $Ba_{0.7}Sr_{0.3}TiO_3$ thin films were examined using an X-ray diffractometer equipped with Cu $K\alpha$ radiation. The $\theta/2\theta$ scan patterns of [001], [011] and

Target- substrate distance	50 mm
Laser energy	250 mJ
Repetition rate of pulsed laser	10 Hz
Ambient gas	O_2
Total pressure of ambient gas	200 mTorr
Substrate temperature	750 °C
Growth rate	~ 20 nm/min

Table 1. PLD conditions of $Ba_{0.7}Sr_{0.3}TiO_3$ thin films.

[111]-oriented $Ba_{0.7}Sr_{0.3}TiO_3$ films on [001], [011] and [111] MgO substrates, respectively, are shown in Fig. 1 (a)-(c). No secondary orientations and phases can be seen in any of the three XRD patterns, indicating that the BST films are oriented along the particular normal of the substrates with a pure perovskite phase. The full width at half maximum (FWHM) of the x-ray rocking curves (ω scan) for the BST [001], [011] and [111] peaks of the [001], [011] and [111]-oriented BST films are 0.46°, 0.57° and 1.06°, respectively, implying that the crystallites of all three films are fairly well ordered. The in-plane texturing of the BST thin films with respect to the major axes of the MgO substrates was confirmed by the XRD φ scan of the BST [110], [010] and [100] reflections of the [001], [011] and [111]-oriented BST films. The peaks from BST films coincide in position well with those from MgO substrates, as shown in Fig. 1 (e)-(f), which suggests a nonlattice–rotated epitaxial growth of all the as-deposited BST films. For the optical applications, epitaxial growth is strongly desirable because of the basic requirements for the reduction of light scattering associated with the refractive index mismatch at grain boundaries (Lee et al., 1996) as well as the E-O properties comparable to the bulks (Wessels, 2004).

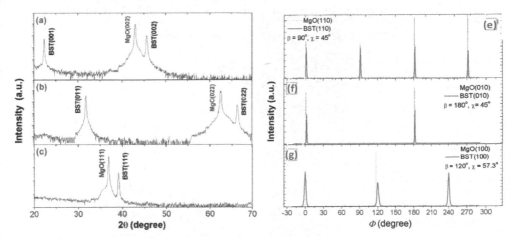

Fig. 1. XRD θ/2θ scan patterns of $Ba_{0.7}Sr_{0.3}TiO_3$ thin films deposited on (a) MgO [001], (b) MgO [011] and (c) MgO [111] single crystal substrates; XRD φ scans of (e) $Ba_{0.7}Sr_{0.3}TiO_3$ [110] and MgO [110] reflections of [001]-oriented film (f) $Ba_{0.7}Sr_{0.3}TiO_3$ [010] and MgO [010] reflections of [011]-oriented film and (f) $Ba_{0.7}Sr_{0.3}TiO_3$ [100] and MgO [100] reflections of [111]-oriented film, providing nonlattice-rotating epitaxial growth of $Ba_{0.7}Sr_{0.3}TiO_3$ thin films. [Reproduced with permission from Ref. (Wang et al., 2010). Copyright 2010, AIP]

3. Optical properties of $Ba_{0.7}Sr_{0.3}TiO_3$ thin films

3.1 Optical transmittance and band gap energy

Optical constants can be evaluated using the "envelop method" developed by (Manifacier et al, 1976). For an insulating film on a transparent substrate, assuming the film is weakly absorbing and the substrate is completely transparent, the optical band gap energy E_{gap} and refractive index n can be derived from the transmission spectra. The optical transmission of the $Ba_{0.7}Sr_{0.3}TiO_3$ thin films was measured using a Perkin Elmer (precisely) Lambda 950 UV-VIS spectrometer in the wavelength range of 200-2000 nm. All three BST films are highly transparent in the visible to near infrared regions, as shown in Fig. 2, which is favorable for applications in optical communication (e. g. λ = 1.3 and 1.5 μm). The transparency of the films drops sharply in the UV region and the threshold wavelength is located at 311, 319 and 317 nm for [001], [011] and [111]-oriented films, respectively. The optical band gap energy E_{gap} of a thin film can be deduced from the spectral dependence of the absorption constant $\alpha(\upsilon)$ by applying the Tauc relation (Tauc, 1972):

$$\alpha h\upsilon = const(h\upsilon - E_{gap})^{1/r} \tag{1}$$

where υ is the frequency and h is the Planck's constant, r = 2 for a direct allowed transition. The absorption constant $\alpha(\upsilon)$ is determined from the transmittance spectrum using the relation (Davis & Mott, 1970):

$$\alpha(\upsilon) = [\ln\frac{1}{T(\upsilon)}] / d \tag{2}$$

where $T(\upsilon)$ is the transmittance at frequency υ and d is the film thickness. Thickness of the BST thin films were measured by alpha-step profiler. (187, 193 and 225 nm for [001], [011]

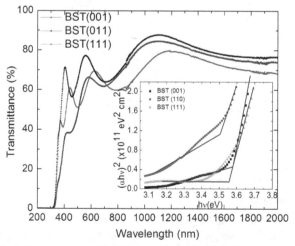

Fig. 2. Optical transmission spectra of [001], [011] and [111]-oriented $Ba_{0.7}Sr_{0.3}TiO_3$ thin films. Inset is the plots of $(\alpha h\upsilon)^2$ versus $h\upsilon$ for $Ba_{0.7}Sr_{0.3}TiO_3$ thin films. The optical band gap energy E_{gap} is deduced from extrapolation of the straight line to $(\alpha h\upsilon)^2 = 0$. [Reproduced with permission from Ref. (Wang et al., 2010). Copyright 2010, AIP]

and [111]-oriented films, respectively). The optical band gap energy is then obtained by applying a "base line" method (Marple, 1966) in order to minimize the impact of reflectance losses at air-film and film-substrate interfaces. Inset of Fig. 2 shows the plots of $(\alpha h\upsilon)^2$ versus $h\upsilon$ for $Ba_{0.7}Sr_{0.3}TiO_3$ thin films grown on MgO substrates. The optical band gap energies are found to be 3.57 ± 0.01, 3.50 ± 0.02 and 3.55 ± 0.01 eV for for the [001], [011] and [111] oriented films, respectively. It is discernable that the absorption edges and optical band gap energy of the $Ba_{0.7}Sr_{0.3}TiO_3$ films are orientation dependent. Although the grain size effect on optical band gap energy of BST thin films has been reported (Thielsch et al.,1997), it is believed that orientation is the predominant factor that is responsible for the observed difference of E_{gap} in our case, as the average crystallites dimensions estimated by Scherrer's formula for the three films are comparable. Similar orientation dependence of band gap energy has also been observed in other oxygen-octahedral perovskite ferroelectrics, such as $Pb(Mg_{1/3}Nb_{2/3})O_3$-$PbTiO_3$ (Wan et al., 2005) etc.

The refractive index n of the as-deposited BST films was derived on the basis of the following expressions (Manifacier et al., 1976),

$$n = \sqrt{N' + \sqrt{N'^2 - n_s^2}} \tag{3}$$

$$N' = \frac{1 + n_s^2}{2} + \frac{2n_s(T_{max} - T_{min})}{T_{max}T_{min}} \tag{4}$$

in which T_{max} and T_{min} are the corresponding maximum and minimum of the envelop around the interference fringes at a certain wavelength λ, n_s is the refractive index of MgO, which is taken from Ref. (Bass, 1994) based on Sellmeier dispersion equation. The experimental values of the refractive index are found to fit closely to a Cauchy function as a formula:

$$n = A + (B / \lambda)^2 + (C / \lambda)^4 \tag{5}$$

where A, B, C are determined from fits to the experimental spectra. The calculated n and the dispersion of the refractive index for the three BST thin films were given in Fig. 3. The dispersion curves rise rapidly towards shorter wavelengths, showing the typical shape of dispersion near an electronic interband transition. Conspicuous orientation dependence of refractive index is discernable, especially in near infrared region. The [001]-oriented film exhibits the highest refractive index in near IR. The optical properties of an oxygen-octahedral ABO_3 perovskite ferroelectric are dominated by BO_6 octahedra, which govern the low-lying conduction bands and the highest valence bands. This lowest energy oscillator is the largest contributor to the dispersion of the refractive index (Chan et al., 2004). Meanwhile, voids caused by surface roughness and porosity inside the film is another controlling factor for the variation of refractive index (Chan et al., 2004; Yang et al., 2002). Moreover, changes in electronic structure due to lattice distortion and some variations of atomic coordination caused by the substrate orientations (Tian et al., 2002) may also be responsible for the observed variation in refractive index of differently oriented BST thin films.

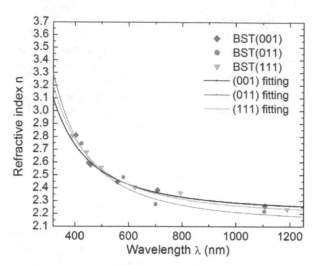

Fig. 3. Variation of refractive indices of [001], [011] and [111]-oriented $Ba_{0.7}Sr_{0.3}TiO_3$ thin films as a function of wavelength. The solid lines are the fitting curves based on Cauchy equation. [Reproduced with permission from Ref. (Wang et al., 2010). Copyright 2010, AIP]

3.2 Electro-Optic properties

Although the E-O properties of bulk ferroelectric materials, such as single crystals of $LiNbO_3$ and $BaTiO_3$, and PLZT transparent ceramics, are well identified, the identification of the E-O properties of ferroelectric thin films is not easy since they exhibit processing-dependent properties (Nashimoto et al., 1999) and crystallographic anisotropy (Wang et al., 2010). The E-O properties of the $Ba_{0.7}Sr_{0.3}TiO_3$ thin films were measured with a transverse geometry at the wavelength of 632.8 nm using modified Sénarmont method. The electrode pattern used for E-O characterization consisted of two coplanar electrodes, with dimensions 1.0×8.0 mm^2 separated by a 20-μm-wide gap. The experimental arrangement for E-O measurement is illustrated in Fig. 4. The film surface was set with the direction perpendicular to the incident light and the electric field was applied normal to the incident light beam. The light beam from a 2 mW stabilized He-Ne laser, after passing through a polarizer set at -45°, impinged normally on the film in the gap between two gold electrodes. The laser beam was then modulated at 50 kHz by the PEM-90 photoelastic modulator and then passed through an analyzer set at + 45°. The transmitted laser beam was detected by a photomultiplier tube (PMT). The electrical signal from the PMT was filtered by a band-pass filter and then fed to a SRS SR830 DSP lock-in amplifier. An electric field was then applied. The general expression for the light intensity I at the detector is given by:

$$I = 1 - \cos(B)\cos(A) + \sin(B)\sin(A)$$

(6)

where B is the phase retardation in the film sample, $A = A_0\cos(\Omega t)$ is the phase retardation in PEM-90 photoelastic modulator. A Fourier series expansion yields:

$$I = [1 - \cos(B)J_0(A_0)] + 2\sin(B)J_1(A_0)\cos(\Omega t) + 2\cos(B)J_2(A_0)\cos(2\Omega t) + \ldots$$

(7)

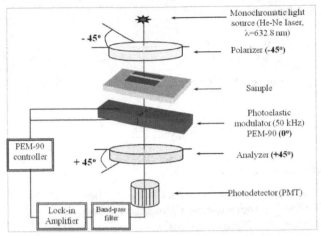

Fig. 4. Schematic diagram showing the modified Sénarmont method for the measuring E-O coefficients of thin films. [Reproduced with permission from Ref. (Wang et al., 2007a). Copyright 2007, AIP]

where $J_n(A_0)$ are the Bessel functions and the first, second and third terms represent the DC term, the fundamental term and the first harmonic term, respectively. Therefore, the electrical signals corresponding to these three terms are:

$$V_{DC} = G[1 - \cos(B)J_0(A_0)] \tag{8}$$

$$V_{1(peak)} = 2G\sin(B)J_1(A_0)\cos(\Omega t) \tag{9}$$

$$V_{2(peak)} = 2G\cos(B)J_2(A_0)\cos(2\Omega t) \tag{10}$$

where G is a constant of proportionality. If A_0 is chosen such that $J_0(A_0)=0$, then the DC signal is independent of the sample retardation B. This occurs for $A_0=2.405$ radians. The DC signal may therefore be used to "normalize" the fundamental term:

$$\frac{V_{1(peak)}}{V_{DC}} = 2\sin(B)J_1(A_0) \tag{11}$$

In Eq. (6), $V_{1(peak)}$ is the peak voltage of the signal. However, the lock-in amplifier only gives the rms voltage V_{rms}. For sinusoidal waveforms, V_{rms} is given by:

$$V_{rms} = \frac{V_{(peak)}}{\sqrt{2}} \tag{12}$$

If we define a ratio R_{1f} as:

$$R_{1f} = \frac{V_{1(rms)}}{V_{DC}} = \sqrt{2}\sin(B)J_1(A_0) \tag{13}$$

then the phase retardation B is given by:

$$B = \sin^{-1}[\frac{R_{1f}}{\sqrt{2}J_1(A_0)}] = \sin^{-1}[\frac{R_{1f}}{\sqrt{2}J_1(2.405)}] = \sin^{-1}[\frac{R_{1f}}{\sqrt{2}\cdot(0.5191)}] \qquad (14)$$

Then the electric field induced birefringence change $\delta(\Delta n)$ can be deduced from the phase change B:

$$\delta(\Delta n) = \frac{\lambda B}{2\pi \cdot d} \qquad (15)$$

where d is the thickness of the film.

The field induced birefringence of the thin films was characterized as a function of d. c. electric field E at room temperature and the result is shown in Fig. 5, in which strong orientation dependence of E-O effect is clearly seen. E-O effect for the [011]-oriented film is relatively weak, while large birefringence changes $\delta\Delta n$ are revealed in [001] and [111]-oriented films. All the three films exhibit predominantly linear birefringence change with respect to the applied d. c. electric field. A slight hysteresis behavior is observed in the $\delta\Delta n$ versus E plots for [001]-oriented film, which is consistent with its enhanced ferroelectric properties (Wang et al., 2005). Generally, the birefringence shift due to linear E-O effect (Pockels effect) is given by Eq. (16) as following,

$$\delta\Delta n = \frac{1}{2}n^3 r_c E \qquad (16)$$

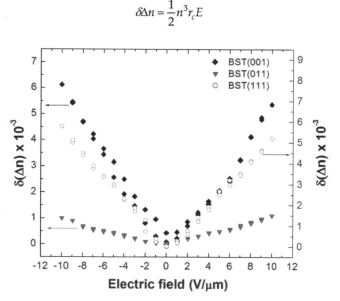

Fig. 5. Change in birefringence $\delta(\Delta n)$ as a function of applied d.c. electric field for [001], [011] and [111]-oriented $Ba_{0.7}Sr_{0.3}TiO_3$ thin films. The birefringence was determined at a wavelength of 632.8 nm. [Reproduced with permission from Ref. (Wang et al., 2010). Copyright 2010, AIP]

With this Eq. (16), the effective linear electro-optic coefficient r_c can be deduced from the slope of the $\delta\Delta n$ versus E plots. In this case, the linear E-O coefficient r_c of the [001], [011] and [111]-oriented BST thin films were calculated to be 99.1 pm/V, 15.7 pm/V and 87.8 pm/V, respectively. The difference in E-O properties in the three kinds of oriented BST films may be attributed to the changes in distribution and magnitude of spontaneous polarization (electric) in orientation engineered films. The polarization changes could originate from: (1) the magnitude variation of the relative displacement of the Ti^{4+} with respect to O^{2-} in the octahedral structure, (2) the change of domain growth mechanism, and (3) the lattice distortion caused by the stain in perovskite structure (Lu et al., 1999; Moon et al., 2003). Other factors, such as dielectric permittivity, may also be responsible for the orientation dependence of E-O effect in our tetragonal-distorted BST thin films (Wan et al., 2004). Nevertheless, the linear E-O coefficients r_c of [001] and [111]-oriented BST films are considerable higher than that of commonly used LiNbO$_3$ single crystals (30.8 pm/V) (Xu, 1991), showing their potential for use in active waveguide applications. The understanding of orientation dependent optical properties of BST thin films is technically important for practical optoelectronic device development. Due to its largest E-O coefficient, we will focus on [001]-oriented BST thin films in the following sections.

3.3 Light propagation characteristics

To characterize the waveguide properties of the Ba$_{0.7}$Sr$_{0.3}$TiO$_3$ films, prism coupling experiments were performed at wavelengths of both 632.8 nm and 1550 nm. Fig. 6 shows the guided mode spectra (m-lines) of a 620 nm thick Ba$_{0.7}$Sr$_{0.3}$TiO$_3$ thin film on MgO [001]

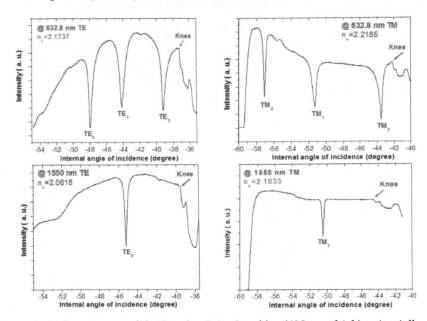

Fig. 6. Prism coupling spectra of the Ba$_{0.7}$Sr$_{0.3}$TiO$_3$ thin film (620 nm thick) epitaxially grown on MgO [001] substrate at both 632.8 nm and 1550 nm. [Reproduced with permission from Ref. (Wang et al., 2006a). Copyright 2006, OSA]

substrate measured by the prism coupler. Three TE (transverse electric) and three TM (transverse magnetic) modes were observed at 632.8 nm, while only a single TE and TM mode were found at 1550 nm. The peaks of each guided mode are very sharp and distinguishable, indicating that a good confinement of light propagation is achieved and the film is potentially useful for optical waveguide devices. The measured film thickness of 626.0 nm in the TE mode at 632.8 nm was in good agreement with that determined by an α-step profile measurement. At the wavelength of 632.8 nm, the refractive indices for TE and TM modes were determined to be 2.1696 and 2.2185, respectively, giving an index difference of 0.0489. The large index difference cannot be fully explained using the intrinsic birefringence in the film. Perhaps, this phenomenon is attributed to the strain induced by the lattice mismatching. The well-defined and relatively sharpen m-lines suggest that the optical losses in the film are rather low since the optical losses are related to the FWHM of the m-lines (Dogheche et al.; 2003Vilquin et al., 2003). It is obvious that the FWHM of m-lines at 1550 nm is smaller than that at 632.8 nm, a lower optical loss is expected at 1550 nm.

From the knowledge of the effective mode indices, it is possible to determine the refractive index profile along the film thickness direction for either TE or TM modes by using the inverse Wentzel-Kramers-Brillouin (i-WKB) method (Chiang, 1985). This method only depends on the refractive index distribution within the guiding layer. The refractive index profile of the $Ba_{0.7}Sr_{0.3}TiO_3$ thin film at 632.8 nm is shown in Fig. 7. It indicates a step-like index variation, which is synonymous with a good optical homogeneity along the BST film thickness. The refractive index remains constant within the guiding region and drops rapidly near the film-substrate interface. It is because plenty of lattice misfit induced dislocations exist near the film-substrate interface (Chen et al., 2002), thus degrading the optical properties at the interface.

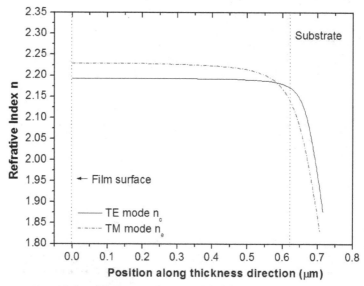

Fig. 7. Reconstruction of the refractive index profile of the $Ba_{0.7}Sr_{0.3}TiO_3$ thin film (620 nm thick) epitaxially grown on MgO [001] substrate at 632.8 nm using an i-WKB method. [Reproduced with permission from Ref. (Wang et al., 2006a). Copyright 2006, OSA]

3.4 Optical loss

For integration, the films are not only required to have E-O properties comparable to those of the bulk but also must have a high degree of microstructure perfection in order to minimize optical scattering losses. In this sense, the optical transparency is a demanding requirement for thin film waveguides and the optical propagation loss has been the most serious barrier for practical applications of ferroelectric thin films in the waveguide devices. Loss of about 2 dB/cm will reduce the efficiency of an optical device (e. g. frequency doublers) by over 50% (Fork et al., 1995). Nevertheless, investigations of the optical losses in ferroelectric thin films are insufficient. The optical loss is mainly caused by absorption, mode leakage, internal scattering and surface scattering. For a transparent ferroelectric thin film, the dominant loss mechanism is the scattering (Lu et al., 1998). In this study, a "moving fibre method" (build-in option of Metricon 2010 prism coupler) was employed to determine the surface scattering losses in the $Ba_{0.7}Sr_{0.3}TiO_3$ thin film (620 nm thick). The measurement setup is shown schematically in Fig. 8. In the moving fibre method, the exponential decay of light is measured by a fibre probe scanning down the length of the propagating streak. The optical fibre method is identical in concept to the CCD camera approach for measuring the decay of the propagating streak as described in previous work (Lu et al., 1998; Walker et al., 1994).

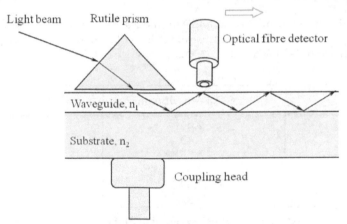

Fig. 8. A schematic diagram of the experimental arrangement for surface scattering loss measurements (moving fibre method). [Reproduced with permission from Ref. (Wang et al., 2006a). Copyright 2006, OSA].

The loss was derived for the film from measurements of the out-of-plane scattered light intensity for the specified guiding modes. Fig. 9 shows the scattered intensity from TE_0 and TM_0 modes at both 632.8 nm and 1550 nm. A least square fit gives the losses of 2.64 dB/cm and 3.04 dB/cm for TE_0 and TM_0 modes at 632.8 nm, respectively and 0.93 dB/cm and 1.29 dB/cm for TE_0 and TM_0 modes at 1550 nm, respectively. Losses with similar magnitude were measured for the other guided modes of the $Ba_{0.7}Sr_{0.3}TiO_3$ thin film as summarized in Table 2. For the modes of higher order, higher scattered losses were observed. It is noticeable that the scattered losses at 1550 nm for the commonly used wavelength in optical communication are rather low, which is consistent with our prediction using m-line measurements. The accuracy of our loss results is limited due to the small sample area and thus the short scanning length along the light propagation direction. But it gives a good approximation and the results are comparable with previous reported data in ferroelectric thin films (Beckers et al., 1998; Lu et al., 1998; Walker et al., 1994).

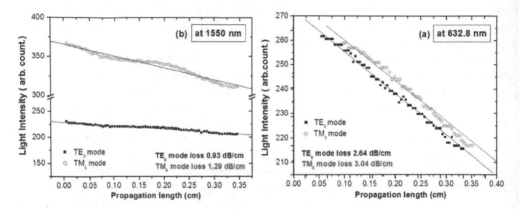

Fig. 9. Scattered intensity from the TE_0 and TM_0 modes of a 620 nm thick $Ba_{0.7}Sr_{0.3}TiO_3$ thin film epitaxially grown on MgO [001] substrate at both 632.8 nm and 1550 nm. [Reproduced with permission from Ref. (Wang et al., 2006a). Copyright 2006, OSA]

Losses Guided mode	at 632.8 nm (dB/cm)	at 1550 nm(dB/cm)
TE_0	2.64	0.93
TE_1	6.43	--
TE_2	8.33	--
TM_0	3.04	1.29
TM_1	5.39	--
TM_2	8.85	--

Table 2. Surface scattered losses in $Ba_{0.7}Sr_{0.3}TiO_3$ thin film epitaxially grown on MgO [001] substrate.

4. Waveguide device design

For device speed and efficiency, the ideal waveguide structure should consist of a thin film material with a large E-O coefficient deposited onto a substrate possessing a small microwave dielectric constant (Lu et al., 1998). The thin films should have low optical loss and low surface roughness. Moreover, for better light confinement, a large refractive index difference between the substrate and the film is desired. Therefore, the BST/MgO configuration is a favorite structure for use in waveguide applications. In this work, the effective index method (Dogheche et al., 1996; Kogelnik & Ramaswamy, 1974; Ramaswamy, 1974) was employed to study the bidimensional waveguide. The physical structure of a BST/MgO ridge waveguide was schematically shown in Fig. 10.

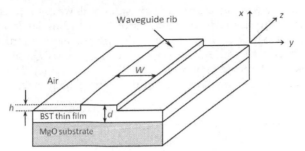

Fig. 10. Schematic structure of BST/MgO rib waveguide. [Reproduced with permission from Ref. (Wang et al., 2006b). Copyright 2006b, Elsevier]

For guide-wave applications, the design of a bidimensional waveguide begins with the optimization of the geometrical parameters of a planar waveguide. Consider the case of an asymmetric planar waveguide of an anisotropic active film with an isotropic cladding layer and a buffer layer. Since the film is highly oriented with the [001] direction parallel to the x-direction and the grain orientations in the y and z directions are completely random, we can assume that the average index in both y and z direction are the same, i.e., $n_y = n_z$, and the waveguide is therefore uniaxial. The dispersion equations for an asymmetric uniaxial planar thin film waveguide with an isotropic cover layer are given by:

$$kd\sqrt{n_y^2 - N_m^2} = m\pi + \sum_{i=1,2} \tan^{-1}(\sqrt{\frac{N_m^2 - n_i^2}{n_z^2 - N_m^2}}) \tag{17}$$

for the TE polarization and by

$$kd\sqrt{n_x^2 - N_m^2} = m\pi + \sum_{i=1,2} \tan^{-1}(\frac{n_y}{n_i})^2(\sqrt{\frac{N_m^2 - n_i^2}{n_x^2 - N_m^2}}) \tag{18}$$

for the TM polarization. Where N_m is the effective index of the mode with the mode number $m = 1, 2, 3 \ldots$, n_i is the index of the ith layer, d is the film thickness and k (= $2\pi/\lambda$) is the free space wave number. To realize the bidimensional structure, the first step is to study the effect of the film thickness d on guided modes propagation. Fig. 11 shows the dispersion of the effective index N_m as a function of the guiding layer thickness d. For the planar guiding structure, we have studied the transverse confinement of the light. Therefore, we should consider only the TE polarization modes. In order to minimize the waveguide losses, it may be designed such that the transverse resonance condition could be satisfied; in order to allow only the TE₀ guided mode to propagate along the active layer. As shown in Fig. 11, the number of guided modes increases as the film thickness increases. At a wavelength of 1550 nm, the cutoff of TE₀ and TE₁ modes are 0.19 µm and 0.86 µm, respectively. In our study, a film thickness of 620 nm corresponds to a single-mode propagation at 1550 nm, resulting in the transverse confinement of the mode.

To establish both the horizontal and vertical guiding confinement of single propagation mode, the geometry of the rib (width W and height h) must be determined for a given film thickness d (=620 nm). For lateral confinement of the light, the evolution of effective index as a function of the width W was calculated. We consider a rib waveguide as shown in Fig. 12 (a), where n_c, n_f and n_s are the refractive indices of the cladding layer, film and substrate with $n_f > n_c$, n_s. The basic idea is to replace the rib waveguide structure by a fictive equivalent planar waveguide with the effective indices N_{eff1} and N_{eff2} obtained from planar waveguides of thickness d and $(d-h)$ as shown in Fig. 12 (b) and (c). Then the problem has been simplified to solving the "thickness" W in the lateral planar waveguide shown in Fig. 12 (c) for a given rib height h.

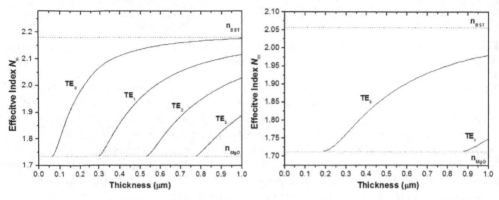

Fig. 11. Dispersion of effective index N_m of the modes versus the film thickness d for a BST/MgO structure. (a) $\lambda = 632.8$ nm, (b) $\lambda = 1550$ nm. [Reproduced with permission from Ref. (Wang et al., 2006b). Copyright 2006, Elsevier]

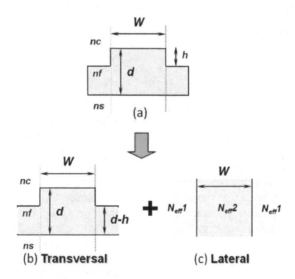

Fig. 12. Description of the effective index method for a rib waveguide. (a) the original rib waveguide; (b) solving the vertical slab problem to define N_{eff1} and N_{eff2}; (c) solving the equivalent horizontal slab problem to determine the N_{eff} of the whole structure. [Reproduced with permission from Ref. (Wang et al., 2006b). Copyright 2006, Elsevier]

Fig. 13 shows the simulation results of the evolution of the effective index as a function of the rib width W for some given rib height h at both 633 nm and 1550 nm, which gives a good description of the influence of rib geometry on the effective index in considering a lateral single-mode propagation. The useful region of the single-mode rib waveguide is a function of the geometrical parameters of the rib (W and h). The cut-off width of the TE_{01} mode depends on the value of height h. At $\lambda=1550$ nm, for $h = 50$ nm in Figure 6 (b), the waveguide becomes multimode for W greater than 2.4 µm. Actually, for a deeper height h, the width W has to be narrower to maintain a single-mode lateral propagation. Meanwhile, a larger rib height may cause serious scattering loss if optical scattering from the sidewall of the waveguide is significant. Although the accuracy of cut-offs calculated by effective index method needs to be improved, it does provide a good reference and approximation for us to shape the waveguide geometry far from the cut-offs in order to achieve a single-mode rib waveguide.

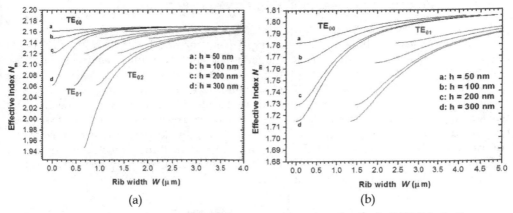

Fig. 13. Effective index N_m versus rib width W for various heights h of a BST/MgO rib waveguide. (a) $\lambda = 632.8$ nm, (b) $\lambda = 1550$ nm. [Reproduced with permission from Ref. (Wang et al., 2006b). Copyright 2006, Elsevier]

In integrated optics, interferometric devices such as optical sensors and electro-optic amplitude modulators are often of the Mach-Zehnder interferometer (MZI) type (Krijnen et al., 1995). The proposed structure of the $Ba_{0.7}Sr_{0.3}TiO_3$ Mach-Zehnder modulator investigated in this work is shown schematically in Fig. 14. The modulator consists of an input waveguide that branches out into two separate parallel waveguides that are finally recombined into the output waveguide. By applying a voltage to the coplanar electrodes, the phase of the guided light in this branch can be changed. All the waveguides in the modulator are of the rib type and the geometry of the rib is chosen based on the numerical calculation in order to achieve single mode propagation at a wavelength of 1550 nm. The active arm length and the device length are set at 5000 µm and 9000 µm, respectively, to assure smooth curvatures at the Y-branches. In practice, the branching angle is usually kept below 1º in order to reduce the insertion loss because the loss increases with branching angle (Agrawal, 2004). However, due to the limited length (1 cm) of our thin film samples, the branching angle is set at 3º.

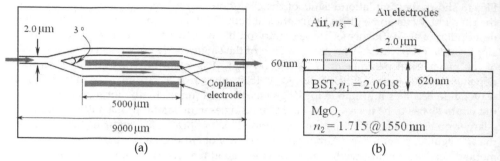

(a)   (b)

Fig. 14. Geometry of the investigated BST/MgO Mach-Zehnder E-O modulator. (a) Planar view; (b) Cross-sectional view of the active arm.

5. Waveguide device processing

To fabricate a rib waveguide from BST thin film, the process includes standard photolithographic patterning and dry etching. The microfabrication flow chart is shown in Fig. 15. After cleaning the film surface, a 150 nm thick chromium film was deposited on the $Ba_{0.7}Sr_{0.3}TiO_3$ film by rf magnetron sputtering to serve as the etch barrier layer. A positive photoresist was then deposited on the chromium layer by spin coating, resulting in a photoresist layer of 1 μm thickness. After baking, the photoresist layer was exposed under a mask to high intensity ultraviolet light in a mask aligner. The exposed photoresist was then immersed in a developer to release the desired waveguide pattern. Using an etching solution of $Ce(NO_3)_4$ in aceric acid, the chromium layer without the protection of photoresist was removed, leaving the bare waveguide patterns of the undeveloped photoresist covering the chromium barrier layer on the film surface.

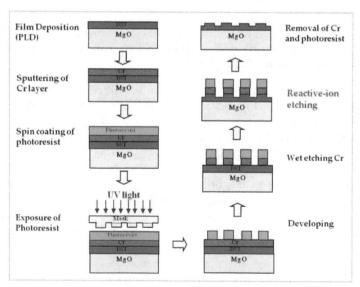

Fig. 15. Microfabrication flow chart of thin film rib waveguides.

Dry etch of the BST layer was performed using reactive ion etching in a CF_4/Ar ambient. A rf power of 200 W was used and the gas pressure was kept at 4.7 Pa. A 30-min etching results in ~ 60 nm depth as measured using an alpha-step surface profiler. Fig. 16 shows the SEM image of a cleaved BST/MgO rib waveguide. It is clearly seen that geometry of the rib is 620 nm of thickness, 60 nm of etched depth and 1.9 μm of width. These dimensions meet the requirements for a single-mode ridge waveguide at 1550 nm based on the theoretical calculation so that a single-mode propagation along the rib is expected. For Mach-Zehnder modulators, Au top electrode layer of 150 nm thick was deposited by rf magnetron sputtering, followed by a photolithographic patterning and wet chemical etching. The coplanar electrodes length and gap were 5.0 mm and 10 μm, respectively. The modulator was oriented along the BST [110] direction.

Fig. 16. SEM image of a cleaved BST/MgO rib waveguide. [Reproduced with permission from Ref. (Wang et al., 2006b). Copyright 2006, Elsevier]

6. Characterization of rib waveguide and Mach-Zehnder modulator

The near-field output pattern of the BST/MgO rib waveguide was measured using an end-fire coupling technique (Wang et al., 2006b). Fig. 17 shows a schematic diagram of the end-fire coupling method. Light source is a TE-polarized semiconductor laser with a wavelength of 1550 nm. A polarization-maintaining single mode fibre was used for input and butt-coupled to the cleaved endface of the waveguide. A charge-coupled device (CCD) camera was used to image the output pattern of the waveguide through a micro-objective lens. Fig. 18 shows the near-field output pattern of the BST/MgO rib waveguide. It illustrates that the output intensity is a single mode beam with Gaussian beam profile, showing that a strong light beam is propagating along the rib structure.

The Mach-Zehnder modulator was first tested without applying electric field using the end-fire coupling method at a wavelength of 1550 nm to confirm its single propagation mode characteristics. The E-O response (modulation of light intensity) of the $Ba_{0.7}Sr_{0.3}TiO_3$ Mach-Zehnder modulator was characterized using the setup shown in Fig. 19 (Wang et al., 2007b). A light beam (λ = 1550 nm) from a laser diode was coupled into a single-mode optical fibre. A fibre polarizer was used for defining the polarization state of the input light (TE). The light beam was end-fire coupled into the Mach-Zehnder modulator. The output of the waveguide was butt-coupled into another single mode fibre, which was fed to an IR photodetector and then to an optical spectrum analyzer. Two xyz micro-positioning systems were used to position the optical fibres at the input and output of the modulator. Contact needles, supported by micromanipulators, were used to apply the voltage to the coplanar electrodes. The transmitted optical intensity was recorded by an oscilloscope. The half-wave voltage V_π was determined to be 60 V by applying dc bias field to the modulator. To evaluate the electro-optic modulation under an ac electric field, a triangular voltage with a frequency of 25 Hz and peak voltage of 120 V was applied to the Mach-Zehnder modulator.

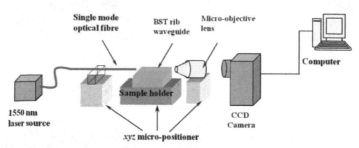

Fig. 17. Schematic diagram showing the measurement of the near-field output pattern by the end-fire coupling method. [Reproduced with permission from Ref. (Wang et al., 2006b). Copyright 2006, Elsevier]

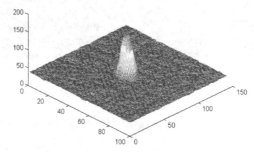

Fig. 18. The output light intensity exhibits a Gaussian profile, showing that only a single TE_{00} mode propagates along the BST/MgO rib waveguide at a wavelength of 1550 nm. [Reproduced with permission from Ref. (Wang et al., 2006b). Copyright 2006, Elsevier]

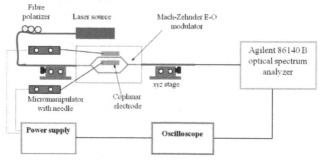

Fig. 19. Experimental setup for the characterization of Mach-Zehnder E-O modulator. [Reproduced with permission from Ref. (Wang et al., 2007b). Copyright 2007, Taylor & Francis]

The device exhibited good response to the applied ac voltage as shown in Fig. 20. A frequency doubling phenomenon was observed in the output signal as compared with the driving voltage. The performance of the device was very stable up to a frequency of 1 MHz. The E-O coefficient of the Ba$_{0.7}$Sr$_{0.3}$TiO$_3$ thin film Mach-Zehnder modulator can be calculated from the half-wave voltage V_π. Since Ba$_{0.7}$Sr$_{0.3}$TiO$_3$ thin film grown on MgO [001] substrate exhibits the linear E-O effect, the phase change B in the device can be derived from Eqs. (19) and (20):

$$B = \frac{2\pi L}{\lambda}\delta(\Delta n) = \frac{2\pi L}{\lambda}\cdot(\frac{1}{2}n^3 r_c^{eff} E) = \frac{\pi \cdot n^3 LV}{\lambda s}r_c^{eff} \tag{19}$$

where L is the activation length of the device, s is the coplanar electrode gap spacing, λ the wavelength of the light, V the applied voltage, and r_c^{eff} is the effective E-O coefficient. Putting $B = \pi$,

$$r_c^{eff} = \frac{\lambda \cdot s}{n^3 LV_\pi} \tag{20}$$

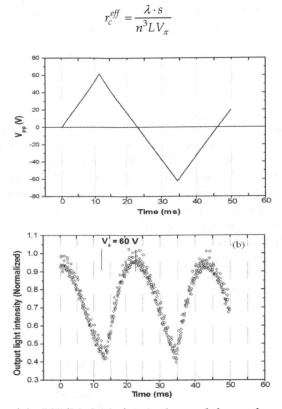

Fig. 20. E-O response of the BST/MgO Mach-Zehnder modulator when an ac voltage is applied. Upper trace is the applied 25 Hz trianglar driving voltage on 5.0-mm-long electrode; Bottom trace is the intensity modulation output signal at a wavelength of 1550 nm. [Reproduced with permission from Ref. (Wang et al., 2007b). Copyright 2007, Taylor & Francis]

Therefore, the effective E-O coefficient of the device at $\lambda = 1550$ nm is calculated to be $r_c^{eff} =$ 27.0×10^{-12} m/V, which is comparable to that obtained for $LiNbO_3$ E-O modulators (Yariv & Yeh, 1983).

7. Conclusions

In summary, we studied the optical and electro-optic properties of ferroelectric $Ba_{0.7}Sr_{0.3}TiO_3$ thin films grown on MgO single-crystal substrates by pulsed laser deposition. XRD examination confirmed an epitaxial growth and a pure perovskite phase with good single crystal quality. The thin films possess excellent optical clarity with high optical transmission in visible light – near infrared regime. E-O properties of the $Ba_{0.7}Sr_{0.3}TiO_3$ thin films were measured at a wavelength of 632.8 nm through a modified Sénarmont method. A strong correlation between the optical, electro-optic properties and the crystalline orientation of epitaxial $Ba_{0.7}Sr_{0.3}TiO_3$ thin films were revealed. The linear electro-optic coefficient r_c of the [001], [011] and [111]-oriented thin films were found to be 99.1 pm/V, 15.7 pm/V and 87.8 pm/V, respectively. Understanding of the optical and electro-optic anisotropy in ferroelectric thin films is a critical issue for the device design and fabrication. Because of its largest E-O coefficient, the [001]-oriented $Ba_{0.7}Sr_{0.3}TiO_3$ thin films have been identified to be the most promising candidate for the integrated optics applications.

Waveguide characteristics of the [001]-oriented $Ba_{0.7}Sr_{0.3}TiO_3$ thin films were determined using prism coupling technique. The films showed good optical homogeneity along the thickness direction by analyzing the guided mode spectra using the i-WKB method. The films exhibited relatively low surface scattered losses of 0.93 dB/cm and 1.29 dB/cm for TE_0 and TM_0 modes at 1550 nm, respectively, which is very favorite for use in infrared waveguides. Appropriate geometry of the rib waveguide structure was computed using the effective index method. A rib waveguide based on $Ba_{0.7}Sr_{0.3}TiO_3$ thin film grown on MgO substrate was successfully fabricated by photolithographic patterning and dry etching. A single mode (TE_{00}) propagation along the rib was observed, which agrees well with the numerical calculation. Thin-film Mach-Zehnder waveguide modulator from the $Ba_{0.7}Sr_{0.3}TiO_3$/MgO heterostructures has also been demonstrated. The measured half-wave voltage V_π is 60 V and the effective E-O coefficient r_c^{eff} of this device is calculated to be 27.0 $\times 10^{-12}$ m/V at a wavelength of 1550 nm. Our results show that BST thin film optical modulator is attractive promising candidate for the practical applications in optical communications.

8. Acknowledgment

Financial supports from the Vice-Chancellor's postdoctoral fellowship program, The University of New South Wales (SIR50/PS16940) and Australian Research Council Discovery Project (Grant No. DP110104629) are acknowledged.

9. References

Adachi, H. & Wasa, K. (1991). Sputtering preparation of ferroelectric PLZT thin films and their optical applications. *IEEE Trans.Ultra. Ferro. Freq. Contr.* 38: 645-655.

Agrawal, G. P. (May 2004). *Lightwave Technology: Components and Devices*. John Wiley & Sons, ISBN: 978-0-471-21573-8, New York.

Bass, M. (September 1, 1994). *Handbook of Optics, Vol. 2: Devices, Measurements, and Properties Second Edition, McGraw-Hill Professional*, ISBN-13: 978-0070479746, New York.

Beckers, L.; Schubert, J.; Zander, W.; Ziesmann, J.; Echau, A.; Leinenbach P. & Buchal, Ch. (1998). Structural and optical characterization of epitaxial waveguiding BaTiO₃ thin films on MgO. *J. Appl. Phys.* 83: 3305-3310.

Blomqvist, M.; Khartsev, S. & Grishin, A. (2005). Electrooptic ferroelectric Na$_{0.5}$K$_{0.5}$NbO₃ films. *IEEE Photonics Tech. Lett.* 17:1638-1640.

Chan, K. Y.; Tsang, W. S.; Mak C. L.; Wong, K. H. & Hui, P. M. (2004). Effect of composition of PbTiO₃ on optical properties of (1-x)PbMg$_{1/3}$Nb$_{2/3}$O$_{3-x}$-PbTiO₃ thin films. *Phys. Rev. B.* 69: 144111.

Chen, C. L.; Garret, T.; Lin, Y.; Jiang, J. C.; Meletis, E. I.; Miranda, F. A.; Zhang, Z. & Chu, W. K. (2002). Interface structures and epitaxial behavior of ferroelectric (Ba, Sr)TiO₃ thin films. *Integr. Ferro.* 42: 165-172.

Chen, D.; Hafich, M.; Huber, J. & Kyonka, J. (1980). PLZT modulators for optical communications. *Ferroelectrics*, 27: 73-76.

Chiang, Kin. S. (1985). Construction of refractive index profiles of planar dielectric waveguides from the distribution of effective indexes. *J. Lightwave. Tech.* LT-3: 385-391.

Davis, E. A. & Mott, N. F. (1970). Conduction in non-crystalline systems. 5. Conductivity, optical absorption and photoconductivity in amorphous semiconductors. *Phil. Mag.* 22: 903.

Dogheche, E.; Lansiaux, X. & Remiens, D. (2003). *m*-line spectroscopy for optical analysis of thick LiNbO₃ layers grown on sapphire substrates by radio-frequency multistep sputtering. *J. Appl. Phys.* 93: 1165-1168.

Dogheche, E.; Rémiens, D. & Thierry, B. (1996). Optimum parameters in the design of electrooptic waveguide modulators using ferroelectric thin films. *Proc. 10th IEEE Int'l. Sym. Appl. Ferro.* 1: 61-64.

Fork, D. K.; Armani-Leplingard, F. & Kingston, J. J. (1995). Optical losses in ferroelectric oxide thin films: is there light at the end of the tunnel? *Mater. Res. Soc. Symp. Proc.* 361: 155-166.

Gopalakrishnan, G. K.; Burns, W. K.; McElhanon, R. W.; Bulmer, C. H.; Greenblatt, A. S. (1994). Performance and modeling of broadband LiNbO₃ travelling wave optical intensity modulators. *J. Lightwave Technol.* 12:1807–1819.

Graettinger, T. M.; Rou, S. H.; Ameen, M. S.; Auciello, O. & Kingon, A. I. (1991). Electro-optic characterization of ion beam sputter-deposited KNbO3 thin films. *Appl. Phys. Lett.* 58: 1964-1966.

Guarino, A.; Poberaj, G.; Rezzonico, D.; Degl'innocenti, R. & Günter, P. (2007). Electro-optically tunable microring resonators in lithium niobate. *Nature Photonics.* 1: 407-410.

Haertling, G. H. (1987). PLZT electrooptic materials and applications-a review. *Ferroelectrics*, 75: 25-55.

Hoerman, B. H.; Nichols, B. M. & Wessels, B. W. (2003). The electro-optic properties of epitaxial KTa$_x$Nb$_{1-x}$O₃ thin films. *Opt. Commun.* 219: 377-382.

Kang, T. D.; Xiao, B.; Avrutin, V.; Özgür, Ü.; Morkoç, H.; Park, J. W.; Lee, H. S.; Lee, H.; Wang, X. Y. & Smith, D. J. (2008). Large electro-optic effect in single-crystal Pb(Zr, Ti)O$_3$ (001) measured by spectroscopic ellipsometry. *J. Appl. Phys.* 104: 093103.

Kim, D. H. & Kwok, H. S. (1995). Pulsed laser deposition of BaTiO$_3$ thin films and their optical properties. *Appl. Phys. Lett.* 67: 1803-1805.

Kim, D. Y.; Moon, S. E.; Kim, E. K.; Lee, S. J.; Choi, J. J. & Kim, H. E. (2003). Electro-optic characteristics of (001)-oriented Ba$_{0.6}$Sr$_{0.4}$TiO$_3$ thin films. *Appl. Phys. Lett.* 82: 1455-1457.

Kogelnik, H. & Ramaswamy, V. (1974). Scaling rules for thin-film optical waveguides. *Appl. Opt.* 13: 1857-1862.

Krijnen, G. J. M.; Villeneuve, A.; Stegeman, G. I.; Aitchison, S.; Lambeck, P. & Hoekstra, J. W. M. (1995). Modelling of a versatile all-optical Mach-Zehnder switch. In: *Guided-Wave Optoelectronics: Device Characterization, Analysis, and Design*. Tamir, T.; Griffel, G. & Bertoni, H. F. pp. 187-196, Springer, ISBN-13: 978-0306451072, New York.

Lee, S. H.; Noh, T. W. & Lee, J. H. (1996). Control of epitaxial growth of pulsed laser deposited LiNbO$_3$ films and their electro-optic effects. *Appl. Phys. Lett.* 68: 472-474.

Li, J. W.; Duewer, F.; Gao, C.; Chang, H.; Xiang, X. D. & Lu, Y. L. (2000). Electro-optic measurements of the ferroelectric-paraelectric boundary in Ba$_{1-x}$Sr$_x$TiO$_3$ materials chips. *Appl. Phys. Lett.* 76: 769-771.

Lin, P. T.; Liu, Z. & Wessels, B. W. (2009). Ferroelectric thin film photonic crystal waveguide and its electro-optic properties. *J. Opt. A: Pure Appl. Opt.* 11: 075005.

Lu, Y. L.; Jin, G.-H.; Golomb, M. C.; Liu, S.-W.; Jiang, H.; Wang, F.-L.; Zhao, J.; Wang, S. –Q. & Drehman, A. J. (1998). Fabrication and optical characterization of Pb(Mg$_{1/3}$Nb$_{2/3}$)O$_3$-PbTiO$_3$ planar thin film optical waveguides. *Appl. Phys. Lett.* 72: 2927-2929.

Lu, Y. L.; Zheng, J. J.; Golomb, M. C.; Wang, F. L.; Jiang, H. & Zhao, J. (1999). In-plane electro-optic anisotropy of (1-x)Pb(Mg$_{1/3}$Nb$_{2/3}$)O$_3$-xPbTiO$_3$ thin films grown on (001)-cut LaAlO$_3$. *Appl. Phys. Lett.* 74: 3764-3766.

Manifacier, J. C.; Gasiot, J. & Fillard, J. P. (1976). A simple method for the determination of the optical constants n, k and the thickness of a weakly absorbing thin film. J. Phys. E: Sci. Instrum. 9: 1002-1004.

Marple, D. T. F. (1966). Optical absorption edge in CdTe: experimental. *Phys. Rev.* 150: 728-734.

Masuda, S.; Seki, A. & Masuda, Y. (2010). Influence of crystal phases on electro-optic properties of epitaxially grown lanthanum-modified lead zirconate titanate films. *Appl. Phys. Lett.* 96: 072901.

Masuda, S.; Seki, A.; Shiota, K. & Masuda, Y. (2011). Mach-Zehnder interferometer-type photonic switches based on epitaxially grown lanthanum-modified lead zirconate titanate films. *J. Lightwave Technol.* 29: 209-213.

Moon, S. E.; Kim, E. K.; Kwak, M. H.; Ryu, H. C.; Kim, Y. T.; Kang, K. Y.; Lee, S. J. &. Kim, W. J. (2003). Orientation dependent microwave dielectric properties of ferroelectric Ba$_{1-x}$Sr$_x$TiO$_3$ thin films. *Appl. Phys. Lett.* 83: 2166-2168.

Nakada, M.; Shimizu, T.; Miyazaki, H.; Tsuda, H.; Akedo, J. & Ohashi, K. (2009). Lanthanum-modified lead zirconate titanate electro-optic modulators fabricated using aerosol deposition for LSI interconnects. *Jpn. J. Appl. Phys.* 48: 09KA06.

Nashimoto, K.; Nakamura, S.; Morikawa, T.; Moriyama, H.; Watanabe, M. & Osakabe, E. (1999). Electrooptical properties of heterostructure (Pb, La)(Zr, Ti)O₃ waveguides on Nb–SrTiO₃. *Jpn. J. Appl. Phys.* 38: 5641-5645.

Petraru, A.; Schubert, J.; Schmid, M. & Buchal, Ch. (2002). Ferroelectric BaTiO₃ thin-film optical waveguide modulators. *Appl. Phys. Lett.* 81: 1375-1377.

Ramaswamy, V. (1974). Propagation in asymmetrical anisotropic film waveguides. *Appl. Opt.* 13: 1363-1371.

Suzuki, M.; Nagata, K. & Yokoyama, S. (2008). Imprint properties of optical Mach-Zehnder interferometers using (Ba, Sr)TiO₃ sputter-deposited at 450 °C. *Jpn. J. Appl. Phys.* 47: 2879-2901.

Tang, P. S.; Meier, A. L.; Towner, D. J. & Wessels, B. W. (2005). BaTiO₃ thin-film waveguide modulator with a low voltage–length product at near-infrared wavelengths of 0.98 and 1.55 μm. *Opt. Lett.* 30: 254-256.

Tauc, J. (1972). Optical properties of non-crystalline solids, In: *Optical Properties of Solids, Abeles, F. pp. 277, North-Holland,* ISBN-13: 978-0444100580, *Amsterdam.*

Tayebati, P.; Trivedi, D. & Tabat, M. (1996). Pulsed laser deposition of SBN:75 thin films with electro-optic coefficient of 844 pm/V. *Appl. Phys. Lett.* 69: 1023-1025.

Thielsch, R.; Kaemmer, K.; Holzapfel, B. & Schultz, L. (1997). Structure-related optical properties of laser-deposited Ba$_x$Sr$_{1-x}$TiO₃ thin films grown on MgO (001) substrates. *Thin Solid Films.* 301: 203-210.

Tian, H. Y.; Choi, J.; No, K.; Luo, W. G. & Ding, A. L. (2002). Effect of compositionally graded configuration on the optical properties of Ba$_x$Sr$_{1-x}$TiO₃ thin films derived from a solution deposition route. *Mater. Chem. Phys.* 78: 138-143.

Uchiyama, K.; Kasamatsu, A.; Otani, Y. & Shiosaki, T. (2007). Electro-optic properties of lanthanum-modified lead zirconate titanate thin films epitaxially grown by the advanced sol-gel method. *Jpn. J. Appl. Phys.* 46: L244-246.

Vilquin, B.; Bouregba, R.; Poullain, G.; Murray, H.; Dogheche, E. & Remiens, D. (2003). Crystallographic and optical properties of epitaxial Pb(Zr$_{0.6}$,Ti$_{0.4}$)O₃ thin films grown on LaAlO₃ substrates. *J. Appl. Phys.* 94: 5167-5171.

Walker, F. J.; McKee, R. A.; Yen, Huan-wun. & Zelmon, D. E. (1994). Optical clarity and waveguide performance of thin film perovskites on MgO. *Appl. Phys. Lett.* 65: 1495-1497.

Wan, X. M.; Luo, H. S.; Zhao, X. Y.; Wang, D. Y.; Chan, H. L. W. & Choy, C. L. (2004). Refractive indices and linear electro-optic properties of (1-x)Pb(Mg$_{1/3}$Nb$_{2/3}$)O₃-xPbTiO₃ single crystals. *Appl. Phys. Lett.* 85: 5233-5235.

Wan, X. M.; Zhao, X. Y.; Chan, H. L. W.; Choy, C. L. & Luo, H. S. (2005). Crystal orientation dependence of the optical band gap of (1-x)Pb(Mg$_{1/3}$Nb$_{2/3}$)O$_{3-x}$-PbTiO₃ single crystals. *Mater. Sci. Phys.* 92: 123-127.

Wang, D. Y.; Wang, Y.; Zhou, X. Y.; Chan, H. L. W. & Choy, C. L. (2005). Enhanced in-plane ferroelectricity in Ba$_{0.7}$Sr$_{0.3}$TiO₃ thin films grown on MgO (001) single-crystal substrate. *Appl. Phys. Lett.* 86: 212904.

Wang, D. Y.; Chan, H. L. W. & Choy, C. L. (2006a). Fabrication and characterization of epitaxial Ba$_{0.7}$Sr$_{0.3}$TiO₃ thin films for optical waveguide applications. *Appl. Opt.* 45: 1972-1978.

Wang, D. Y.; Lor, K. P.; Chung, K. K.; Chan, H. P.; Chiang, K. S.; Chan, H. L. W. & Choy, C. L. (2006b). Optical Rib Waveguide Based on Epitaxial $Ba_{0.7}Sr_{0.3}TiO_3$ Thin Film Grown on MgO, *Thin Solid Films*. 510: 329-333.

Wang, D. Y.; Wang, J.; Chan, H. L. W. & Choy, C. L. (2007a). Structural and electro-optic properties of $Ba_{0.7}Sr_{0.3}TiO_3$ thin films grown on various substrates using pulsed laser deposition. *J. Appl. Phys.* 101: 043515.

Wang, D. Y.; Lor, K. P.; Chung, K. K.; Chan, H. P.; Chiang, K. S.; Chan, H. L. W. & Choy, C. L. (2007b). Mach-Zehnder electro-optic modulator based on epitaxial $Ba_{0.7}Sr_{0.3}TiO_3$ thin films, *Ferroelectrics*, 357: 109-114.

Wang, D. Y.; Li, S.; Chan, H. L. W. & Choy, C. L. (2010). Optical and electro-optic anisotropy of epitaxial $Ba_{0.7}Sr_{0.3}TiO_3$ thin films. *Appl. Phys. Lett.* 96, 061905.

Wessles, B. W.; Nystrom, M. J.; Chen, J.; Studebaker, D. & Marks, T. J. (1996). Epitaxial niobate thin films and their nonlinear optical properties. *Mat. Res. Soc. Symp. Proc.* 401: 211-218.

Wessels, B. W. (2004). Thin Film Ferroelectrics for Guided Wave Devices. *J. Electroceramics.* 13: 135-138.

Wessels, B. W. (2007). Ferroelectric epitaxial thin films for integrated optics. *Annu. Rev. Mater. Res.* 37: 659-679.

Wooten, E. L.; Kissa, K. M.; Yi-Yan, A.; Murphy, E. J.; Lafaw, D. A.; Hallemeier, P. F.; Maack, D.; Attanasio, D. V.; Fritz, D. J.; McBrien, G. J. & Bossi, D. E. (2000). A review of lithium niobate modulators for fiber optic communication systems. *IEEE J. Sel. Top. Quant. Electron.* 6:69–82.

Xu, Y. (November 1, 1991). *Ferroelectric Materials and Their Applications*. North-Holland, ISBN-13: 978-0444883544, Amsterdam.

Yang, S. H.; Mo, D.; Tian, H. Y.; Luo, W.G.; Pu, X. H. & Ding, A. L. (2002). Spectroscopic ellipsometry of $Ba_xSr_{1-x}TiO_3$ thin films prepared by the sol–gel method. *Phys. Stat. Sol. (a).* 191: 605-612.

Yariv, A. & Yeh, P. (1983). *Optical waves in crystals: propagation and control of laser radiation*. John Wiley & Sons, ISBN 13: 9780471091424, New York.

Zhu, M. M.; Du, Z. H. & Ma, J. (2010). Influence of crystal phase and transparent substrate on electro-optic properties of lead zirconate titanate films. *J. Appl. Phys.* 108: 113119.

8

Electrical Control of Nonlinear TM Modes in Cylindrical Nematic Waveguide

Carlos G. Avendaño[1], J. Adrian Reyes[2] and Ismael Molina[3]
[1,3]Autonomous University of Mexico City, G. A. Madero, Mexico D. F.,
[2]Institute of Physics, National Autonomous University of Mexico, Mexico D. F.,
Mexico

1. Introduction

Liquid crystals (LCs) are intermediate phases between the solid and liquid states of matter whose interesting properties are owing mainly to two remarkable characteristics: i) they can flow as a conventional liquid, ii) they possess positional and orientational order just like those of the solid crystals (de Gennes & Prost, 1993). During the last five decades, LCs have been widely used in optoelectronical devices due to the great ability of changing their properties under the stimuli of external agents as temperature, pressure and electromagnetic fields. It is well known that the propagation of an electromagnetic wave through LCs is a phenomenon that exhibit unique optical properties and highly nonlinear effects (Zel'dovich et al., 1980; Tabiryan et al.,1986).

It is an experimentally well established fact that a polarized and sufficiently intense laser beam may distort the initial orientation of a liquid crystal sample reorienting its molecules against the elastic torques producing a new equilibrium orientational configuration. This orientational transition of the same mesophase is the so called optical Freedericksz transition (de Gennes & Prost, 1993). For pure LCs this phenomenon occurs for linear, circular or elliptically polarized beams and, in the reorientation process, different nonlinear dynamical regimes may be achieved (Durbin et al., 1981). The understanding of the underlying physical mechanisms and the prediction of the ensuing changes in the optical properties of the liquid crystals is an active area of research nowadays (Khoo & Wu, 1993; Santamato et al., 1990).

LCs are anisotropic materials and their linear optical properties are described by a symmetric dielectric tensor, instead of a scalar refractive index. Nonetheless, for liquid crystal films where an uniform orientation is achieved, the dielectric tensor is constant and light propagation through the fluid may be described by the usual laws of crystal optics (Born & Wolf , 1975). But for spatially inhomogeneous liquid crystal layers light propagation is much more difficult to describe, essentially due to the fact that there is no general method to solve Maxwell's equations for an arbitrary spatial dependence of the dielectric tensor. However, for important special cases such as the optical phenomena observed in the cholesteric phase, exact solutions and useful approximations have been worked out if light propagates along the helical axis (Oseen, 1933). For light propagation in an arbitrary direction relative to the helix, the description is more difficult. For this situation Berreman and Scheffer developed a numerical method to solve Maxwell's equations. This method can be applied to any system where the director changes only in one direction, i. e., for a stratified medium (Berreman & Scheffer, 1970; Shelton & Shen, 1972). But since numerical methods give little insight into the physical features of the problem, approximate solutions of Maxwell's equations have been developed mostly in the context of light propagation in cholesterics. One of these cases is the geometrical optics approximation for LC. This approximation has been formulated in terms of the concept of adiabatic propagation for an arbitrary stratified medium (Allia et al., 1987), or for the case of normal incidence and small birefringence (Santamato & Shen, 1987). On the other hand, a rigorous treatment of the geometrical optics approximation in the special case of a stratified layer with its director oriented everywhere in the plane of incidence of the beam, was presented by Ong (Ong, 1987). But apparently, the generalization of the adiabatic or geometrical optics approximations have not been extended for two or three dimensional spatial variations of the director.

When a high intensity beam is propagated in LCs whose configuration is not anchored to waveguide boundary conditions, give rise to spatial patterns and solitons as a result of the balance between the nonlinear refraction and the spatial diffraction. It is shown that for nematic LCs the electromagnetic field amplitude at the center of Gaussian beam (inner solution), follows a nonlocal nonlinear Schrödinger equation (McLaughlin et al., 1996). For cholesteric LCs and wavelengths outside of the bandgap, it is found that under special conditions the nonlinear coupled equations for the wavepackets in the sample reduces to an extended nonlinear Schödinger equation with space-dependent coefficients (Avendaño & Reyes, 2004), whereas for wavelengths within the bandgap (stationary waves) the vectorial equation reduces to an extended real Ginzburg-Landau equation (Avendaño & Reyes, 2006). In this system the energy exchanging among the four different modes generated in the sample due to linear and nonlinear coupling is also studied.

It is worth mentioning that the analyses made in the above cited works, the nonlinear effects are obtained in regions of the system where both orientational and optical field have lost influence from the boundary conditions and they have to satisfy only certain mean-field matching conditions. Indeed, as long as the confining cell of the liquid crystal turns to be larger, the bias-free confinement is more notorious.

If the boundary conditions are to be considered, the study of transverse magnetic (TM) nonlinear modes in nematic LC core waveguides can be realized by two different assumptions: i) by assuming hard anchoring boundary conditions for the nematic director, an iterative numerical scheme permits determine up to certain approaches the propagation constant as a function of optical power (Lin& Palffy-Muhoray, 1994), ii) by considering soft anchoring boundary conditions, a numerical but exact procedure allows to obtain the propagating parameters, transverse field distribution and nematic configuration as a function of the mode intensity (Avendaño & Reyes, 2010). It is shown that the anisotropy of the nematic and the intensity of the propagating beam causes simultaneously spatial redistribution of the field amplitude and the nematic configuration, as well as changes in the propagation constant and on the cut off frequencies. As said above, LCs change their properties under external stimuli, so that, it is expected that any external agent will permit us to control these nonlinear parameters.

In next section we review some aspects of the propagation of light in inhomogeneous nematic liquid crystal waveguide consisting of an isotropic core and a quiescent nematic liquid crystal cladding. To this end an analytic and iterative solution of the nematodynamic equations coupled to Maxwell's equations describing the propagation of a narrow wavepacket, is provided. To cubic order in the coupling between the optical field and the non-stationary reorientational states of the nematic, a perturbed Nonlinear Schrödinger Equation (NLS) is derived. This envelope equation that takes into account the dissipative effects due to the presence of hydrodynamic flow in a cylindrical fiber whose nematic cladding is initially quiescent, and the dissipation associated with the reorientation are also analyzed.

In last section we are focussed in analyzing the effect of applying an axial uniform electric field E^{dc} on the nonlinear TM modes, the propagating parameters and nematic core configuration by assuming soft anchoring boundary conditions within a cylindrical waveguide made of a nematic liquid crystal core and isotropic cladding. In order to achieve this goal, Maxwell equations are written for the proposed system and their corresponding boundary conditions. Then, we establish the set of nonlinear coupled equations governing the nematic configuration and the transverse field distribution by including the arbitrary anchoring conditions under the action of the uniform electric field applied axially. After this, we solve numerically the coupled nematic-electromagnetic field system and find simultaneously the distorted textures of the nematic inside the cylinder and the nonlinear TM modes as a function of E^{dc}. We show that the correlation in the spatial distribution of nematic's configuration and nonlinear TM modes, the nonlinear cut-off frequencies and dispersion relations can be tuned by varying the external electric field E^{dc}.

2. Liquid crystal cladding waveguide

We first consider a cylindrical geometry for an optical fiber that takes into account the nonlocal features of the reorientation dynamics. In what follows the coupled time evolution equation for both, the Transverse Magnetic TM modes and for the orientational configuration are derived in an explicit retarded form in terms of the coupling parameter q, which it will be defined later.

Then, these general equations are solved to linear order in q for the final stationary orientational configuration and are then used to construct the propagation equation of a wavepacket of TM modes. It is shown that the envelope of the wavepacket obeys a NLS equation which balances self-focussing, dispersion and diffraction in the nematic. For the soliton solution we calculate its speed, time and length scales, and nonlinear index of refraction. They are estimated by using experimental values for some of the parameters (Chen & Chen , 1994).

2.1 Coupled dynamics

Let us consider a cylindrical waveguide with an isotropic core of radius a, dielectric constant ε_c and a quiescent nematic liquid crystal cladding of radius b satisfying planar axial boundary hard-anchoring conditions $\hat{n}(r = a, z) = e_z$.

The nematic director is written in terms of the angle θ as follows

$$\hat{n}(r,z) = \sin\theta\hat{e}_r + \cos\theta\hat{e}_z , \tag{1}$$

where $\hat{e}_r$ and $\hat{e}_z$ are the unit cylindrical vectors along the r and z directions, respectively. If the reorientation process is isothermal, the equilibrium orientational configurations are determined by minimizing the corresponding total Helmholtz free energy (Frank, 1958)

$$
\begin{aligned}
F &= (1/2)\int dV \left[K_1(\nabla \cdot \hat{n})^2 + K_2(\hat{n}\cdot\nabla\times\hat{n}+v)^2 + K_3(\hat{n}\times\nabla\times\hat{n})^2 - \mathbf{D}(\mathbf{r},t)\cdot\mathbf{E}(\mathbf{r},t) \right] \\
&= (1/2)\int dV \left[K(\nabla\cdot\hat{n})^2 + K(\nabla\times\hat{n})^2 - \mathbf{E}^t(\mathbf{r},t)\cdot\int^t dt'\nabla\times\mathbf{H}^*(\mathbf{r}',t') + \mathbf{E}^a(\mathbf{r},t)\cdot\int^t dt'\nabla\times\mathbf{H}^*(\mathbf{r}',t') \right],
\end{aligned} \tag{2}
$$

where v is the chirality that we take null for a nematic. K_1, K_2 and K_3 are the splay, twist and bend constants of deformation. Here $K = K_1 = K_2 = K_3$ is the elastic constant in the equal elastic constant approximation and the asterisk denotes complex conjugation. Here we have used the constitutive relation $\mathbf{D}(\mathbf{r},\omega) = \varepsilon(\mathbf{r},\omega)\cdot\mathbf{E}(\mathbf{r},\omega)$ with $\varepsilon(\mathbf{r},\omega) = \varepsilon_0\left(\epsilon_\perp(\omega)I + \epsilon_a(\omega)\hat{n}\hat{n}\right)$, where ε_0 is the permitivity of the vacuum, $\varepsilon_a = \varepsilon_{||} - \varepsilon_\perp$ is the dielectric anisotropy, whereas that $\varepsilon_\perp$ and $\varepsilon_{||}$ are perpendicular and parallel dielectric constants to the optical axis, respectively, and which leads to the retarded relation between $\mathbf{E}$ and $\mathbf{D}$ given by

$$
\mathbf{E}(\mathbf{r},t) = \int^t dt'' \epsilon\left(\mathbf{r},t-t''\right)\cdot\mathbf{D}(\mathbf{r},t'') = \mathbf{E}^t(\mathbf{r},t) + \mathbf{E}^a(\mathbf{r},t), \tag{3}
$$

where $\mathbf{E}^t$ and $\mathbf{E}^a$ are electric fields defined by the following nonlocal and retarded relations

$$
\mathbf{E}^t(\mathbf{r},t) = \frac{1}{\varepsilon_0}\int dt'\int dt''\frac{\nabla\times\mathbf{H}(\mathbf{r}',t')}{\epsilon_\perp(t''-t')}, \qquad \mathbf{E}^a(\mathbf{r},t) = \frac{1}{\varepsilon_0}\int dt'\int dt''\frac{\epsilon_a(t''-t')\hat{n}\hat{n}\cdot\nabla\times\mathbf{H}(\mathbf{r}',t')}{\epsilon_{||}\epsilon_\perp}. \tag{4}
$$

In Eqs. (2), (3) and (4) we have substituted $\mathbf{D}(\mathbf{r},t)$ in terms of $\mathbf{H}(\mathbf{r},\omega)$ by using Ampere-Maxwell's law without sources. For the specific geometry Eq. (2) takes the form

$$
\begin{aligned}
F = (1/2)\int rdr \Bigg[& K\left(\frac{\sin\theta}{r} - r\sin\theta\frac{\partial\theta}{\partial z} + \cos\theta\frac{\partial\theta}{\partial r}\right)^2 + K\left(\sin\theta\frac{\partial\theta}{\partial r} + \cos\theta\frac{\partial\theta}{\partial z}\right)^2 \\
& -\left(E_r^{a*}\int^t dt'\frac{\partial H_\phi}{\partial z} - E_z^{a*}\int^t dt'\frac{1}{r}\frac{\partial(rH_\phi)}{\partial r}\right) + E_r^{i*}\int^t dt'\frac{\partial H_\phi}{\partial z} - E_z^{i*}\int^t dt'\frac{1}{r}\frac{\partial(rH_\phi)}{\partial r} \Bigg],
\end{aligned} \tag{5}
$$

If we now minimize Eq. (5) with respect to θ, we find the following Euler-Lagrange equation

$$
\begin{aligned}
\frac{\delta F}{\delta\theta} = & \frac{\partial^2\theta}{\partial\zeta^2} + \frac{1}{x}\frac{\partial}{\partial x}\left(x\frac{\partial\theta}{\partial x}\right) - \frac{\sin\theta\cos\theta}{x^2} \\
& -q^2\left[\frac{\cos 2\theta}{x}\left(E_r^{a*}\int^t dt'\frac{\partial H_\phi}{\partial\zeta} + E_z^a\int^t dt'\frac{1}{x}\frac{\partial(xH_\phi)}{\partial x}\right) + \frac{\sin 2\theta}{x}\left(-E_r^{i*}\int^t dt'\frac{\partial H_\phi}{\partial z} + E_z^{i*}\int^t dt'\frac{1}{x}\frac{\partial(xH_\phi)}{\partial x}\right)\right] = 0,
\end{aligned} \tag{6}
$$

Where $\zeta = z/a$, $x = r/a$, $H_\phi = H_\phi/(c\varepsilon_0 E_0)$ with $c = 1/(\mu_0\varepsilon_0)^{1/2}$ where μ_0 is the magnetic permeability of free space. $E_i^a = E_i/E_0$, with $i = r,z$, are dimensionless variables and $q^2 = \varepsilon_0 E_0^2 a^2 / K$ is a dimensionless variable representing the ratio between electric energy density and elastic one. Notice that we only use the final stationary state for θ defined by (6) due to the large difference between the time scales of reorientation and of time variations of the optical field. In this section we ignore all effects due to absorption.

Since only the TM components are coupled with the reorientation, we assume that the optical field is a TM whose electric and magnetic component are E_r, E_z and H_ϕ. Thus, H_ϕ is governed in general by the nonlinear, nonlocal and retarded equation obtained by substituting Eqs. (4) into Faraday's law, namely,

$$
\frac{a^2}{c^2}\frac{\partial^2 H_\phi}{\partial t^2} = -\int dt' \frac{\left(\frac{\partial^2 H_\phi}{\partial \zeta^2} + \frac{\partial^2 H_\phi}{\partial x^2}\right)(t-t')}{\varepsilon_\perp(r',t')} + \frac{\partial^2}{\partial t \partial \zeta}\int dt' \frac{\varepsilon_a(t')}{\varepsilon_\perp \varepsilon_\parallel}\left[-\sin^2\theta\frac{\partial H_\phi}{\partial \zeta} + \frac{\sin\theta\cos\theta}{x}\frac{\partial x H_\phi}{\partial x}\right](t-t') \quad (7)
$$
$$
- \frac{\partial^2}{\partial t \partial \zeta}\int dt' \frac{\varepsilon_a(t')}{\varepsilon_\perp \varepsilon_\parallel}\left[-\sin\theta\cos\theta\frac{\partial H_\phi}{\partial \zeta} + \frac{\cos^2\theta}{x}\frac{\partial x H_\phi}{\partial x}\right](t-t')
$$

Eq. (6) and (7) define a set of coupled equations for the nematic and optical field (Garcia et al., 2000). Next we solve them iteratively in the weakly nonlinear regime.

2.2 Linear and weakly nonlinear dynamics

The solution of Eq. (4) to zeroth order in q and satisfying the axial boundary conditions defined above, is $\theta^{(0)} = 0$. Substitution of this solution into Eq. (7) and taking a monochromatic beam of frequency ω, we obtain a linear equation for the zeroth order field $U \equiv H_\phi^{(0)}$ which is given by

$$
\left(\frac{\varepsilon_\perp}{x^2} - \varepsilon_\perp \varepsilon_\parallel \left(\frac{\omega a}{c}\right)^2\right)U - \varepsilon_\parallel \frac{\partial^2 U}{\partial \zeta^2} - \frac{\varepsilon_\perp}{x}\frac{\partial}{\partial x}\left(x\frac{\partial U}{\partial x}\right) = 0 \quad (8)
$$

Solving this equation by the method of separation of variables, its propagating solution is given by

$$
U = e^{-i\beta a\zeta} A_1 K_1\left(x\sqrt{\beta^2 a^2 \frac{\varepsilon_\perp}{\varepsilon_\parallel} - \varepsilon_\parallel\left(\frac{\omega_0 a}{c}\right)^2}\right), \quad (9)
$$

where A_1 is an arbitrary constant to be determined by using the boundary conditions. Here $K_1(x)$ is the modified Bessel function of order 1. On the other hand, the monochromatic expression of $H_\phi^c(r,z)$ in the isotropic dielectric core (Jackson, 1984) finite at the origin is

$$
H_\phi^c = e^{-i\beta a\zeta} B_1 J_1\left(x\sqrt{\beta^2 a^2 - \varepsilon_c\left(\frac{\omega_0 a}{c}\right)^2}\right), \quad (10)
$$

where $J_1(x)$ is the Bessel function of order 1 and B_1 is also an undetermined constant. To find the constants A_1 and B_1, it is necessary to impose the following boundary conditions over H_ϕ and its derivative at the boundary (Jackson, 1984),

$$H_\phi^c\big|_{x=1} = U\big|_{x=1}, dH_\phi^c / dx\big|_{x=1} = \varepsilon_\perp^{-1} dU / dx\big|_{x=1}. \tag{11}$$

Thus, by substituting Eqs. (9) and (10) into Eq. (11) we obtain a transcendental equation for the allowed values of β corresponding to each of the permitted modes in the guide.

To obtain the weakly nonlinear equations for θ and H_ϕ, we perform another iteration to find their next nonvanishing order corrections in q. For this purpose we first insert Eq. (9) into Eq. (6) to obtain

$$x\frac{\partial^2\theta}{\partial\xi^2} + \frac{\partial}{\partial x}\left(x\frac{\partial\theta}{\partial x}\right) - \frac{\sin\theta\cos\theta}{x} - \frac{q^2\varepsilon_a A_2^2}{\pi\varepsilon_\parallel\varepsilon_\perp}e^{-2\gamma x}$$

$$\left[2\beta a\cos 2\theta + \frac{\sin 2\theta}{4x^2\varepsilon_\perp}\left[4x^2\left(\beta^2 a^2\varepsilon_a - \varepsilon_\perp\varepsilon_\parallel\left(\frac{\omega_0 a}{c}\right)^2\right) + \varepsilon_\perp\right] \Big/ \sqrt{\frac{\varepsilon_\parallel}{\varepsilon_\perp}\left(\beta^2 a^2 - \varepsilon_\perp\left(\frac{\omega_0 a}{c}\right)^2\right)}\right] = 0, \tag{12}$$

and look for a solution of the form

$$\theta = \theta^{(0)} + q^2\left|A(\zeta,t)U(x,t)\right|^2\theta^{(1)}(r) + \dots \tag{13}$$

where $A(\zeta,t)$ is a slowly varying function of its arguments. Hence the equation for $\theta^{(1)}$ takes the form

$$x\frac{\partial^2\theta^{(1)}}{\partial x^2} + \frac{\partial\theta^{(1)}}{\partial x} - \theta^{(1)} / x - 2(\varepsilon_a\beta a A_2^2 / \pi\varepsilon_\parallel\varepsilon_\parallel)e^{-2\gamma x} = 0 \tag{14}$$

and its solution satisfying the hard anchoring hometropic boundary conditions $\theta(x=1) = \theta(x=b/a) = 0$ may be written in terms of the exponential integral function; however, the resulting complicated equation can be approximated using the asymptotic expressions of these functions with the result is given by

$$\theta^{(1)}(x,\omega) = \frac{\beta a\varepsilon_a J_1\left[\sqrt{\varepsilon_c\left(\frac{\omega_0}{c}a\right)^2 - (\beta a)^2}\right]}{\pi\varepsilon_\parallel\varepsilon_\perp x(a^2 - b^2)}\left[(a^2 - b^2)e^{\gamma a(1-x)} + (b^2 - (xa)^2) + e^{\gamma(a-b)}a^2(1-x^2)\right]. \tag{15}$$

If we now insert this expression into Eq. (8) and expand the result up to first order in q, we arrive at an equation of the form

$$\hat{L}(\beta,\omega,x)H_\phi + q^2\hat{N}(H_\phi) = 0 \tag{16}$$

where the linear and nonlinear operators $\hat{L}$ and $\hat{N}$ are defined, respectively, by

$$\hat{L}(\beta,\omega,x) = \frac{1}{x^2\varepsilon_\parallel\varepsilon_\perp}\left[-\varepsilon_\perp + x^2\varepsilon_\perp\left[\varepsilon_\perp\left(\frac{\omega_0 a}{c}\right)^2 - (\beta a)^2\right] + x\varepsilon_\perp\frac{\partial}{\partial x} + x^2\varepsilon_\perp\frac{\partial^2}{\partial x^2}\right] \tag{17}$$

and

$$\hat{N} = \frac{\varepsilon_a |A(\zeta)U(x,\omega)|^2}{x\varepsilon_\parallel \varepsilon_\perp} - i\beta a \left[U\theta^{(1)}(x) + 3x\theta^{(1)}(x)\frac{dU}{dx} + Ux\frac{d\theta^{(1)}(x)}{dx} \right] A. \tag{18}$$

2.3 Wavepacket

The explicit Fourier representation of a monochromatic field as the one considered in last section, depends of the frequency as $\delta(\omega - \omega_0)$, where δ is the delta function. This suggests that a narrow wavepacket centered around the frequency ω_0 may be expressed in the form:

$$H_\phi(x,\zeta,t) = \ddot{A}(\omega - \omega_0,\zeta)e^{i\beta(\omega_0)a\zeta} U_\phi(x,\omega_0) + cc., \tag{19}$$

where the function $A(\omega - \omega_0,\zeta)$ characterizes the distribution of frequencies around ω_0. We assume that this distribution has a small dispersion $q = (\omega - \omega_0)/\omega_0$. Thus, if the amplitude $H_\phi(x,\zeta,\omega)$ is expanded in a Taylor series around $\omega = \omega_0$ and the inverse Fourier transform of $H_\phi(x,\zeta,\omega)$ is taken, we arrive at

$$H_\phi(x,\zeta,t) = \frac{1}{2\pi}\sum_{n=0}^{\infty}\frac{1}{n!}\frac{d^n}{d\omega^n}U_\phi(x,\omega_0)\int(\omega-\omega)^n\ddot{A}(\omega-\omega_0,\zeta)e^{-i(\omega-\omega_0)t}d(\omega-\omega)e^{i\beta(\omega_0)a\zeta-i\omega_0 t} + cc. \tag{20}$$

Eq. (20) can be written in the more compact form

$$H_\phi(x,\zeta,t) = e^{i\beta(\omega_0)a\zeta-i\omega_0 t} U_\phi(x,\omega_0 + iq\frac{\partial}{\partial T})A(\Xi,T) + cc., \tag{21}$$

where $A(\Xi,T)$ is the Fourier transform of $A(\omega - \omega_0,\zeta)$ and is a slowly varying function of the variables $\Xi \equiv q\zeta$ and $T \equiv qt$. Due to the coupling between the reorientation and the optical field, it is to be expected that when a monochromatic TM mode propagates along the cell, higher harmonics may be generated. Therefore, we assume that the solution of Eq. (16) can be written as the superposition

$$H_\phi(x,\zeta,t) = e^{i\beta(\omega_0)a\zeta-i\omega_0 t} U_\phi(x,\omega_0 + iq\frac{\partial}{\partial T})A(\Xi,T) + q^2 U^{(1)} + q^3 U^{(2)} + cc. \tag{22}$$

The superindices identify the first, second, ..., harmonics. Note that the presence of the powers of q implies that the contribution of the higher order harmonics are smaller than the dominant term which is itself a small amplitude narrow wavepacket.

To describe the dynamics of the envelope $A(\Xi,T)$ we substitute Eq. (22) into Eq. (16) and identify the Fourier variables $i\beta_a = i\beta_{0a} + q\partial/\partial\Xi_1 + q^2\partial/\partial\Xi_2$ and $-i\omega = -i\omega_0 + iq\partial/\partial T$, in consistency with a narrow wavepacket, and where $Z = q\Xi_1 = q\Xi_2$ are the spatial scales associated with upper harmonic contributions. Expanding the resulting expressions and grouping contributions of the same order in q, we find the following expressions

$$q : \hat{L}(i\beta_0 a, -i\omega_0, x)U_\phi(x,\omega_0)A = 0. \tag{23}$$

$$q^2 : \left(\hat{L}(i\beta_0 a, -i\omega_0, x)\frac{\partial U_\phi(x,\omega_0)}{\partial\omega}\frac{\partial}{\partial T} + U_\phi(x,\omega_0)\left[\hat{L}_2(i\beta_0 a, -i\omega_0)\frac{\partial}{\partial T} + \hat{L}_1(i\beta_0 a, -i\omega_0)\frac{\partial}{\partial\Xi_1} \right] \right)A = \hat{L}U^{(1)}, \tag{24}$$

$$q^3 : \left(-\frac{1}{2}\hat{L}(i\beta_0 a, -i\omega_0, x)\frac{\partial^2 U_\phi(x,\omega_0)}{\partial \omega^2}\frac{\partial^2}{\partial T^2} + i\frac{\partial U_\phi(x,\omega_0)}{\partial \omega}\hat{L}_2(i\beta_0 a, -i\omega_0)\frac{\partial^2}{\partial T^2} \right.$$

$$+\hat{L}_1(i\beta_0 a, -i\omega_0)\frac{\partial^2}{\partial \Xi_1 \partial T} + \frac{U_\phi(x,\omega_0)}{2}\left(\hat{L}_{22}(i\beta_0 a, -i\omega_0, x)\frac{\partial^2}{\partial T^2} + \hat{L}_1(i\beta_0 a, -i\omega_0)\frac{\partial}{\partial \Xi_2} \right. \tag{25}$$

$$\left. \left. +\hat{L}_{12}(i\beta_0 a, -i\omega_0)\frac{\partial^2}{\partial \Xi_1 \partial T} + \frac{\hat{L}_{11}(i\beta_0 a, -i\omega_0)}{2}\frac{\partial^2}{\partial \Xi_1^2} \right) \right) A - \hat{N}(U_\phi(x,\omega_0)A) = \hat{L}U^{(2)},$$

where $L_i(i\beta_0 a, -i\omega_0)$, $i=1,2$ denotes the derivative of $L(i\beta_0 a, -i\omega_0)$ with respect to its first or second argument.

Note Eq. (23) reproduces the usual dispersion relation $L_i(i\beta_0 a, -i\omega_0)U_\phi(x,\omega_0) = 0$. Taking the first and second derivatives of Eq. (23) with respect to ω we obtain an expression that will allow us to simplify Eqs. (24) and (25) to yield

$$\hat{L}_1(i\beta_0 a, -i\omega_0)U_\phi(x,\omega_0)\left(\frac{ad\beta}{d\omega}\frac{\partial}{\partial T} + \frac{\partial}{\partial \Xi_1} \right)A = \hat{L}U^{(1)}. \tag{26}$$

This expression is a linear inhomogeneous equation U_1, whose solution is assured to exist by imposing the so called alternative Fredholm condition (Zwillinger, 1989), which is fulfilled if $LU(r,\omega_0) = 0$ and $U(r,\omega_0) \to 0$ as $r \to \infty$. In our case this condition reads explicitly $< LU^{(1)}, U_\phi >= 0$ and since $< LU^{(1)}, U >\neq 0$, implies that $\left(\frac{ad\beta}{d\omega}\frac{\partial}{\partial T} + \frac{\partial}{\partial \Xi_1} \right)A = 0$, which expresses the fact that up to second order in q the envelope A travels with the group velocity $d\beta / d\omega$.

Similarly by taking the second derivative of Eq. (23), substituting the resulting expression into Eq. (25) together with $\left(\frac{ad\beta}{d\omega}\frac{\partial}{\partial T} + \frac{\partial}{\partial Z_1} \right)A = 0$ it leads to an explicit expression for $LU^{(2)}$ which upon using again the alternative Fredholm condition $< LU^{(2)}, U_\phi >= 0$, we find

$$in_2 A|A|^2 + 2\frac{\partial A}{\partial \Xi_2} + ia\frac{d^2\beta}{d\omega^2}\frac{\partial^2 A}{\partial T^2} = 0, \tag{27}$$

where the dimensionless refraction index $\bar{n}_2 = Kn_2 / \varepsilon_0 a^2$ is given by.

$$\bar{n}_2 = \frac{\frac{\varepsilon_a}{\varepsilon_\parallel}\left[\left\langle \frac{U_\phi^3(x,\omega_0)}{x}\frac{dx\theta^{(1)}}{dx}, U_\phi(x,\omega_0) \right\rangle + 3\left\langle \theta^{(1)}U_\phi^2(x,\omega_0)\frac{dU_\phi(x,\omega_0)}{dx}, U_\phi(x,\omega_0) \right\rangle \right]}{\left\langle U_\phi(x,\omega_0), U_\phi(x,\omega_0) \right\rangle}$$

$$= \frac{1}{4}\varepsilon_a^2 \beta a^3 J_1\left(\frac{a}{c}\sqrt{\varepsilon_c \omega_0^2 - \beta^2 c^2} \right)^3 e^{-\gamma b + 2\gamma a}\frac{ae^{\gamma(a-4b)} - ae^{-3\gamma b} + be^{-\gamma(4a-b)} - be^{-3\gamma a}}{\pi \varepsilon_\parallel^2 \varepsilon_\perp b(a^2 - b^2)(e^{-2\gamma a} - e^{-2\gamma b})}. \tag{28}$$

2.4 Soliton dimensions

Using the above expressions, we calculate the values of the properties of the wavepacket, such as the nonlinear contribution n_2 to the refractive index, its coefficient $d^2\beta / d\omega^2$, the soliton typical length and time scales and its speed.

For typical values of the dielectric permitivities, from (28) we get a set of values of n_2 (Frank, 1958) corresponding to the allowed values of β_a. The value corresponding to $\beta_a = 229.59$ is $n_2^{5CB} = 2.902 \times 10^{-24} (km/V)^2$ which is several orders of magnitude larger than its value for glass, $n_2^{SiO_2} = 1.2 \times 10^{-28} (Km/V)^2$. This shows the existence of the giant optical nonlinearity expected for a liquid crystal (Reyes & Rodriguez, 2000). Another physical quantity is the coefficient $(d^2\beta / d\omega^2)$ of the wavepacket given by the third term of Eq. (27). Using: $n_o = 1 + n_o^0 + g_o^1 / (\omega_2^2 - \omega^2)$ where $n_o^0 = 0.4136$, $\omega_1 = 8.9 \times 10^{15} rad/s$, $\omega_2 = 6.68 \times 10^{15} rad/s$, $g_o^0 = 4.8 \times 10^{30} (rad/s)^2$ and $g_o^2 = 1.66 \times 10^{30} (rad/s)^2$ for 5CB from (Tabiryan et al., 1986), we find that $(d^2 k_{n_o} / d\omega^2)^{5CB} \approx 1.1 \times 10^{-4} ps^2 / Km$. Thus, the width of a picosecond pulse traveling in 5CB in the linear regime is doubled in a distance of 0.1 m; while for glass (SiO_2), $(d^2 k_{n_o} / d\omega^2)^{SiO_2} \approx 1.8 ps^2 / Km$, it is doubled in a distance of 0.5 Km. This is consistent with the fact that liquids are considerably more dispersive than solids. Note that Eq. (27) can be rewritten as the NLS equation: $iA|A|^2 + \partial A / \partial \Xi_2 + i\partial^2 A / \partial T^2 = 0$, by using the dimensionless variables $\Xi \equiv \Xi_2 / Z_0$ and $\bar{T} \equiv T / T_0$, where $A_0 \equiv c\varepsilon_0 E_0$ is the amplitude of the optical pulse. Here $Z_0 \equiv \varepsilon_\perp K / (aA_0^2\varepsilon_a^2)$, $T_0 \equiv (d^2\beta / d\omega^2)2\varepsilon_\perp K / (\varepsilon_a^2 A_0^2 a)$) are the soliton length and time scales. As is well known, the NLS equation admits soliton type solutions given by (Moloney & Newell, 1992)

$$A = 2A_0 \sec\left[\bar{T} - \bar{Z}dk / d\omega Z_0 / T_0\right]e^{ik(\omega_0)Z_0\bar{Z} - i\omega_0\bar{T}T_0}. \tag{29}$$

For a 500mW laser at $\lambda = 0.5\mu m$, with a beam waist of $10\mu m$, the field amplitude is $A_0^2 = 1.9 \times 10^6 V/m$. Then by using the materials values given above, the spatial and temporal scales for the pulse turn out to be $Z_0 = 4.2 \times 10^{-5} m$ and $T_0 = 0.21 \times 10^{-11} s$.

From Eq. (30) we find that the soliton propagates with the speed $\bar{v} = v / c$

$$\bar{v} = (Z_0 / T)d\omega / dk = (n / cA_0)\sqrt{\lambda_0 / 2\pi n_2 d^2\beta / d\omega^2}, \tag{30}$$

which for the chosen values of the parameters yields $v^{nem} = 0.1$, which is one order of magnitude smaller than the speed of light c in vacuum, and roughly has the same value as for glass, $v^{SiO_2} = 2.5 \times 10^{-1}$. The difference between v^{nem} and v^{SiO_2} comes from the product $n_2 d^2 (kn_0) / d\omega^2$ in Eq. (27), which measures the balance between nonlinearity and dispersion.

3. Electrical control of nonlinear TM modes in cylindrical liquid crystal core waveguide

It is important to stress that spatial solitons (Long et al., 2007) found in nonlinear systems are coherent structures formed in regions of the system where both orientational and optical fields have lost influence from the boundary conditions. In this sense, all these balanced and robust profiles of energy, called solitons, are asymptotic solutions which are not to be forced by strict boundary conditions but they have to satisfy only certain mean-field matching conditions. In this section we are interested instead in analyze the role played by the boundary conditions within the optical- orientational non linear coupling of a liquid crystal cylindrical waveguide.

Most of the optical calculations in waveguides have been done by assuming hard anchoring boundary conditions for the nematic director. This is inconsistent with the high intensity of the propagating TM mode since in the cylinder wall the electric force can be stronger than the surface elastic force as has been shown before for this geometry (Corella-Madueño et al., 2008). Moreover, when liquid crystals are confined to small cavities, its effect is found to be significant, particularly when elastic energies imposed by the confining volume compete with molecular anchoring energies (Corella-Madueño & Reyes, 2008). Hence we cannot ignore surface elastic terms compared with both bulk elastic terms and electric bulk contributions.

In this section we analyze the behavior of a LC nematic confined within a cylindrical fiber of uniform dielectric cladding in which a high intensity TM mode is propagating and a transversal uniform electric field is axially applied on the system. Our aim is to discern how its propagating parameters, transverse field distribution and nematic configuration depend on the optical mode intensity and the external field amplitude, by assuming soft anchoring boundary conditions.

3.1 Transverse magnetic field

We assume homeotropic anchoring of the nematic LCs molecules at the cylinder wall. For infinite circular cylinders the symmetry implies that θ only depends on the radial distance r and the director is given by Eq. (1) (see Fig. 1).

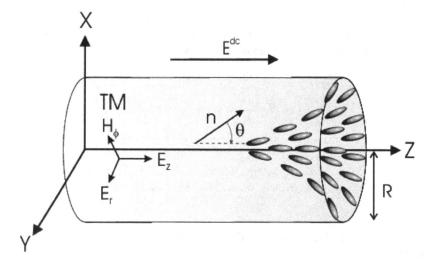

Fig. 1. Cylindrical fiber infiltrated by a nematic liquid crystal and subject to the action of an axial uniform electric field E^{dc} applied along z – axis. Also, an optical field of incident electrical amplitude E_0 is propagating through the sample.

As usual, TM_{lm} and TE_{lm} propagating modes are considered in studying waveguides, nevertheless, as shown in (Lin & Palffy-Muhoray, 1994), for TE_{lm} modes the anisotropy and inhomogeneity of the core does not enter into Maxwell's equations. For these modes the resulting equation is equivalent to that of isotropic and homogeneous cylindrical waveguide (Jackson, 1984). We concentrate on TM_{lm} modes for which the amplitudes of the transverse fields are azimuthally symmetric ($l = 0$). To find the equations governing the propagation of electromagnetic waves through the nematic fiber we assume monochromatic electric E_r, E_z and magnetic H_ϕ fields propagating along the cylinder of the form:

$$(E_r, E_z, H_\phi) = (e_r, e_z, h_\phi)e^{i(\beta z - i\omega t)} \tag{31}$$

where the dimensionless field components are given by the following expression $(e_r, e_z, h_\phi) = E_0(G_r(r, k_0), iG_z(r, k_0), F_\phi(r, k_0)/c)e^{-if}$ and E_0 is the incident electric field amplitude. Here we have explicitly separated the phase f and real valued amplitudes $G_r(r, k_0)$ $F_\phi(r, k_0)$ of the mode components to simplify the resulting equations. Inserting these expressions into Maxwell's equations and separating real and imaginary parts we find (Corella-Madueño & Reyes, 2006):

$$\frac{dG_z}{dx} = -k_0 R \frac{\varepsilon_{rr} - p^2}{\varepsilon_{rr}} F_\phi \tag{32}$$

$$G_z = \frac{1}{k_0 R \varepsilon_\perp \varepsilon_\parallel} \frac{\varepsilon_{rr}}{x} \frac{d(xF_\phi)}{dx} \tag{33}$$

$$G_r = \frac{pF_\phi}{\varepsilon_{rr}} - \frac{i\varepsilon_{rz}}{\varepsilon_{rr}} G_z \tag{34}$$

$$\frac{df}{dx} = \frac{pk_0 R \varepsilon_{rz}}{\varepsilon_{rr}}, \tag{35}$$

where $x \equiv r/R$, R is the cylinder radius and $p \equiv \beta/k_0$, being β the propagation constant. Note that Eqs. (32) and (33) define a self-adjoint equation for F_ϕ so that their eigenvalues p are real, whereas Eq.(35) provides a phase proportional to the only non diagonal entry of ε.

To solve exactly the TM_{0m} modes we shall assume that the nematic cylinder is surrounded by an infinite homogeneous and isotropic cladding of dielectric constant ε_c. In this way the electromagnetic fields should satisfy the boundary conditions analogous to those given by Eqs. (11): $h_\phi(x = 1, k_0) = h_\phi^c(x = 1, k_0)$; $e_z(x = 1, k_0) = e_\phi^c(x = 1, k_0)$ and $h_\phi(x = 0, k_0) = 0$. Where $h_\phi^c(x, k_0)$ and $e_z^c(x, k_0)$ are the magnetic and electric fields in the cladding whose expressions are $h_\phi^c(x, k_0) = AK_1\left(xk_0R\sqrt{p^2 - \varepsilon_c}\right)$ and $e_z^c(x, k_0) = -Ak_0R\sqrt{p^2 - \varepsilon_c}K_0\left(xk_0R\sqrt{p^2 - \varepsilon_c}\right)$, where $K_n(x)$ is the modified Bessel function of order n. Note that, the condition $h_\phi(x = 0, k_0) = 0$ can be derived by realizing that a Frobenius series of the solution of Eq. (32) and (33) has a vanishing independent term. Then, inserting these definitions into boundary conditions, it turns out to be

$$G_z(1) + F_\phi(1)\varepsilon_{\parallel}\varepsilon_{\perp}k_0 R\sqrt{p^2 - \varepsilon_c}\,\frac{K_0\left(k_0 R\sqrt{p^2 - \varepsilon_c}\right)}{K_1\left(k_0 R\sqrt{p^2 - \varepsilon_c}\right)} = 0 \quad \text{and} \quad F_\phi(0) = 0 \qquad (36)$$

The boundary value problem defined by Eqs. (32)-(35), and (36) is twofold: first, it involves coefficients which are real valued functions, and second, it is written in terms of self-adjoint differential operators. Thus, its eigenvalues and eigenfunctions are real.

3.2 Nematic configuration

The continuous medium description of the director is governed by the total free energy F containing the elastic and the optical contributions given by Eq. (2) and the external electric energy after integrating this expression over the cylindrical volume. Then, the free energy per unit length:

$$
\begin{aligned}
F = \pi K_1 &\int_0^1 \left[\left(\frac{d\theta}{dx}\right)^2 (\cos^2\theta + \eta\sin^2\theta) + \frac{\sin^2\theta}{x^2} + \frac{2(K_1 - K_{24})}{K_1} \right] x\, dx \\
&- \pi\varepsilon_{\perp} \int_0^1 \left(|e_z|^2 + |e_r|^2 + |e_\phi|^2 \right) x\, dx \\
&- \frac{\pi q K_1}{E_0^2} \int_0^1 \left[|e_z|^2 \cos^2\theta + |e_r|^2 \sin^2\theta + \frac{\sin 2\theta}{2} \mathrm{Re}\left[e_r e_r^* \right] \right] x\, dx \\
&- \pi\lambda K_1 \int_0^1 \left[\cos^2\theta + \varepsilon_{\perp} / \varepsilon_{\parallel} \right] x\, dx + \pi\sigma K_1 R^2 \cos^2\theta
\end{aligned}
\qquad (37)
$$

Where the elastic moduli K_1, K_2 and K_3 describe the splay, twist and bend deformations, respectively. K_{24} is called the surface elastic constant because it is the coefficient of a divergence term which can be transformed to a surface integral by using Gauss theorem. This elastic constant has to be included because analysis of the Frank free energy for nematics confined to cylindrical regions indicates that the director pattern is dependent on the surface elastic constant K_{24} if there is weak normal anchoring and escape along the cylinder axis (Crawford et al., 1992) $\eta = K_3 / K_1$, $\sigma = RW_\theta / K_1 + K_{24} / K_1 - 1$ and W_θ denotes the strength of interaction between the liquid crystal and the confining surface in units of energy per area. Finally, $q \equiv \varepsilon_a R^2 E_0^2 / K_1$, as seen in section 2, define the ratio between the optical energy and the elastic one; $\lambda \equiv \varepsilon_a R^2 E^{dc2} / K_1$ is another important dimensionless parameter representing the ratio of the external electric and elastic energies; for $\lambda \ll 1$ the influence of the applied field is weak, whereas for $\lambda \gg 1$ the field essentially overcomes the Van der Waals forces between the molecules. To illustrate the order of magnitude of the electromagnetic fields involved, we shall calculate the optical power corresponding to $q = 1$. Let us assume a fiber radius of $R = 10\mu m$. This assures a strong dependence of both texture and electromagnetic fields on the boundary conditions. This leads to an electric amplitude $E_0 = 1.3 \times 10^5 V / m$ which has an irradiance equal to $I = c\varepsilon_{\perp}E_0 / 2 = 2.25 \times 10^7 W / m^2$. If this energy density is distributed across the transverse area of the cylindrical fiber πR^2 we shall obtain a laser power $P = \pi R^2 I = 7 \times 10^{-3} W$. The stationary orientational configuration $\theta(x)$ is determined by minimizing the free energy. This minimization leads to the Euler-Lagrange equation in the bulk

$$0 = \frac{d^2\theta}{dx^2}x^2(\cos^2\theta + \eta\sin^2\theta) + \left(\frac{d\theta}{dx}\right)^2\frac{x^2}{2}(\eta-1)\sin 2\theta + x\frac{d\theta}{dx}(\cos^2\theta + \eta\sin^2\theta) - \frac{\sin 2\theta}{2}$$
$$-q\frac{x^2\sin 2\theta}{2\varepsilon_{rr}^2}\left(-p^2 F_\phi^2 + \varepsilon_\parallel \varepsilon_\perp G_z^2\right) - \lambda\frac{x^2\sin 2\theta}{2} \tag{38}$$

to the condition $\theta(x=0)=0$ in the core and to the arbitrary anchoring boundary condition at the surface

$$d\theta / dx\big|_{x=1} = (\sigma/2)\sin 2\theta / (\cos^2\theta + \sin^2\theta)\big|_{x=1} \tag{39}$$

where we have inserted the conditions Eqs. (33) and (34) in the Euler-Lagrange equation.

3.3 Solutions

We solve this boundary value problem by using the shooting method in which we employ a Runge Kutta algorithm to solve simultaneously Eqs. (32), (33) and (38) by using as initial conditions the right expression of Eq. (36) and arbitrary value for $G_z(0)$ in order to search the value of p and $\alpha = d\theta / dx\big|_{x=0}$ for which the conditions stated in Eqs. (36) and (39) are satisfied. Numerical solutions of Eq. (38) were calculated for 5CB at $T_{IN} - T = 10°C$ with the transition temperature $T_{IN} = 35°C$, $\varepsilon_c = n_c^2 = (1.33)^2$, $\varepsilon_\perp = 2.2201$, $\varepsilon_a = 0.636$, $\eta = 1.316$, $\sigma = 4$, $K_1 = 1.2\times10^{-11}N$, $W_\theta / K_1 = 40\mu m^{-1}$ and $K_{24}/K_1 = 1$ (Crawford et al., 1992). Previous works (Lin & Palffy-Muhoray, 1994) solved separately the electromagnetic boundary problem and the orientational one by following an iterative scheme. Nevertheless, this procedure does not allow to observe the strong correlation in the spatial distribution of nematic's configuration and the transverse modes and hides the dependence of both fields on the optical field intensity, which is related to the parameter q. In addition to this, our procedure permits to observe the influence of the external electric field intensity, which is related to λ, on the optical modes.

3.3.1 Electrical control of *linear* TM modes

Notice that by setting $q=0$ in Eq. (38), we are considering the regime of the *linear* TM optical modes, for which, the textures of the LC are not distorted due to the propagating wave. Then, by varying the parameter λ we have the possibility of controlling electrically the linear TM modes and their propagating parameters.

As it is well known, below a certain values of frequencies ω_c, which are called *cut-off* frequencies, the different optical modes are able to escape from the core and they cannot propagate through the sample. Cut-off frequencies $k_0 R$ as function of the parameter λ are plotted in Fig. 2. Notice that, as λ augments, the cut-off frequencies increases as well. This means that we can electrically control the frequencies for which the modes can be propagated. In fact, any particular propagating mode can be suppressed (or stimulated) by increasing (or decreasing) the external field.

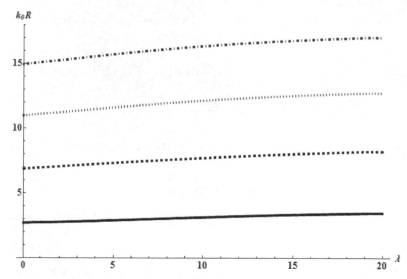

Fig. 2. Cut-off frequencies k_0R for the zeroth (solid line), first (dashed line), second (dotted line) and third (dot-dashed line) modes versus λ.

In Fig. 3 we plot the slope α of the angle θ at the cylinder axis as function of λ. From this graphic, we see that the values of α are degenerated, i.e., they adopt the same value of α for each of the different modes. Additionally, the slope decreases as the external field increases, reaching a limit value, $\alpha = 0$, for values greater than $\lambda \approx 19.3$.

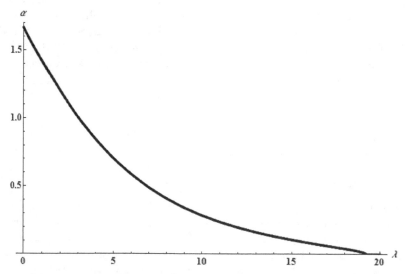

Fig. 3. $\alpha = d\theta(0)/dx$ against λ for the first fourth modes. These modes coincide for each of the values of λ.

As expected, near the axis, the original escaped configuration, $\lambda = 0$, has a higher slope than in the case when the electric field is applied on the waveguide. The effect of the axial electric field E^{dc} on nematic's molecules is to align them along $z-$axis, in such a way, as E^{dc} gets greater, the slope of $\theta(x = 0)$ becomes smaller each time.

Fig. 4 shows the zeroth mode solutions F_ϕ, G_r, G_z and θ as function of the variable x at cut-off frequency for different values of λ. Notice how in general, inside the cylinder, θ diminishes as λ increases, which implies that the effect of electric field over the initial configuration has major effect for soft anchoring than for strong one. This effect is so notorious that, for sufficiently high values of λ, the nematic configuration θ goes to zero for any value of x. This fact agrees with the Fig. 2, for which, the slope α is approximately equal to zero, at the nematic axis, for high electric fields. It is clearly shown that as λ gets larger, the amplitudes of F_ϕ and G_r gets larger as well: in the former case, the maximum amplitude of transverse magnetic field moves to the waveguide axis. This is equivalent to have a higher concentration of energy near the waveguide cladding by augmenting λ.

Finally, the Fig. 5 shows the dispersion relation for the first four modes parametrized by λ. The minimum value of vertical axis takes place at the value $p = n_c = 1.33$ for which the modes cannot propagate, i. e., at $p = n_c$ the corresponding values $k_0 R$ are the cut-off frequencies. Particularly, for $\lambda = 0, 2, 4$, the first mode cut-off frequencies are $k_0 R = 2.7, 2.75, 2.84$, respectively.

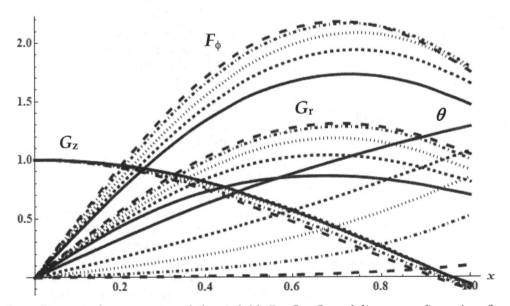

Fig. 4. Dimensionless magnetic and electric fields F_ϕ, G_r, G_z and director configuration θ, at the cut-off frequencies, as function of x for different values of λ: $\lambda = 0$ (solid line), $\lambda = 4$ (dashed line), $\lambda = 8$ (dotted line), $\lambda = 14$ (dot-dashed line) and $\lambda = 19$ (large dashed line).

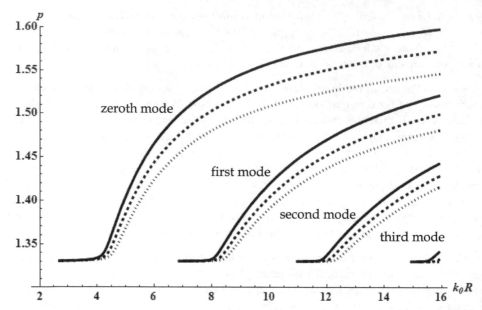

Fig. 5. Dispersion relation p vs k_0R for the first four modes at intensities $\lambda = 0$ (solid line), $\lambda = 2$ (dashed line) and $\lambda = 4$ (dotted line). The minimum value of vertical axis takes places at $p = n_c = 1.33$.

These results imply that the parameter λ plays an important role in controlling the propagating modes. In fact, as λ increases k_0R does as well. Thus, for applications in technology, this external electrical control will permit to design waveguides whose propagating modes can be excited or suppressed by varying the external uniform electric field.

3.3.2 *Nonlinear* TM modes

Nonlinear propagating TM modes can be obtained by arbitrarily increasing the intensity value q. In effect, for values $q > 0$, the nematic configuration given by Eq. (38) depends on the electromagnetic wave amplitude. In this subsection we consider the special case $\lambda = 0$, for which, the electric field E^{dc} is absent.

Fig. 6 shows the cut-off frequencies ω_c against q for the first four modes. As q increases, the cut-off frequencies diminish; and the influence of q on the cut-off frequencies is sharper for smaller q-values. Usually, for frequencies $\omega < \omega_c$, the corresponding TM mode is not propagating. Thus, by enlarging the intensity of the TM mode q, this can be conducted by the guide for lower frequencies than for smaller values of q. However, its influence is reduced when q is larger than certain value and ω_c tends asymptotically to the values shown in this plot. We also notice, by observing Fig. 8, that the influence of q on the configuration of θ is sharper for small values of q, and hence on the cut-off frequencies. It is worth mentioning that, as mentioned in previous section (see Fig. 2), cut-off frequency values gets larger as external electric field λ increases, whereas, cut-off frequencies diminishes as q augments.

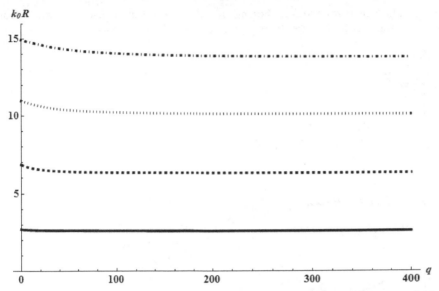

Fig. 6. Cut-off frequencies k_0R for the zeroth (solid line), first (dashed line), second (dotted line) and third (dot-dashed line) modes versus q. As it can be seen, the cut-off frequencies k_0R gets smaller as optical intensity parameter q gets larger.

Fig. 7 shows the slope α of the angle θ at the cylinder axis as function of q. Note that, as expected, when the electromagnetic field is absent, $q = 0$, we obtain only one value for $\alpha = d\theta(0)/dx$, corresponding to the equilibrium configuration of nematic known as escaped configuration. For $q > 0$ the mode amplitude first grows and then decreases against q for different values of the field for each mode. This happens because for small q-values the electric field starts to distort the initial escaped configuration, mostly around $x = 1/2$.

However, once the electric force overcomes the surface elastic force at the cylinder wall $(x = 1)$, the texture is also deformed at the cylinder border and in turn θ is also increased. This causes $\alpha = d\theta(0)/dx$ to diminish since θ is fixed at zero due to the great amount of bulk elastic energy accumulated by the defect of the configuration in the origin.

In Fig. 8 we plot zeroth mode functions $. F_\phi ., G_r , G_z$ and θ as function of the variable x at cut-off frequency for different values of q. As we can see, the maxima of amplitudes of electric field G_r and G_z moves to the cylinder axis, whereas the maximum of amplitude of magnetic field F_ϕ displace to the cylinder border. However, the relative variations of both F_ϕ, G_r and G_z versus q is negligible in comparison with that of θ. This can be understood on the fact that F_ϕ, G_r and G_z fulfill hard boundary conditions whereas θ satisfies soft boundary conditions. In other words, by increasing q, the stationary orientational configuration $\theta(x)$ at the cylinder border gets larger. Particularly, for $q = 400$, the angle $\theta(x)$ ranges from $\theta(x = 0) = 0°$ to $\theta(x = 1) = 90°$, that is, like a homeotropic configuration; while, for $q = 0.001$ the angle θ is approximately $74°$ at the cylinder wall. In other words, as light power increases, the nematic molecules align perpendicular to the cylinder wall. This behavior means that for arbitrary anchoring conditions the field has major effect over the configuration than for hard-anchoring. In addition to this, we see that away from the axis $G_r > G_z$, and the director tends to align in the radial direction as q grows.

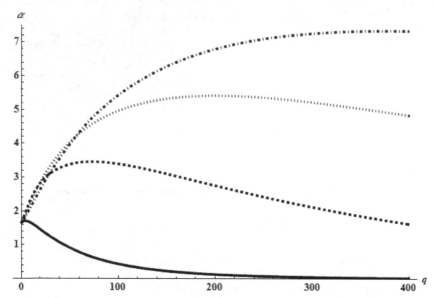

Fig. 7. $\alpha = d\theta(0) / dx$ against q for the same modes of Fig. 6. Notice that, when the electromagnetic field is absent, $q = 0$, we obtain only one value for $\alpha = d\theta(0) / dx$, in agreement with Fig. 3.

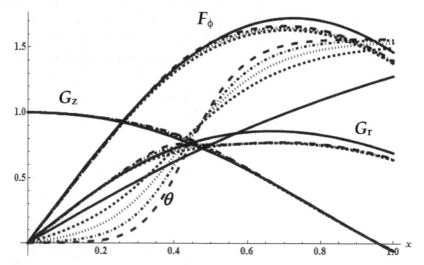

Fig. 8. Dimensionless magnetic and electric fields F_ϕ, G_r, G_z and director configuration θ at the cut-off frequencies, as function of x for five values of q: $q = 0$ (solid line), $q = 50$ (dashed line), $q = 100$ (dotted line), $q = 200$ (dot-dashed line) and $q = 400$ (large dashed line).

Fig. 9 depicts the dispersion relation for the first four nonlinear modes at three different values of q. As expected, for all modes, the minimum value of parameter p is $n_c = 1.33$. This occurs just at cut-off frequencies, for which, the ratio β / k_0 is simply equal to n_c. For particular cases q equal to 0, 5 and 10, the cut-off frequencies $k_0 R$ for the zeroth mode are 2.7, 2.67 and 2.65, respectively. For practical cases, the waveguides are designed so that they can support only the zeroth mode; Fig. 9 can be used for determining some of these useful parameters.

Finally, we mention that, the opposite effect to what we have just said can be seen in Fig. 5, for which the cut-off frequency values gets larger as external electric field λ increases. Therefore, our results show that we can control the propagating or not propagating modes in the waveguide by changing two different parameters: wave amplitude and external electric field.

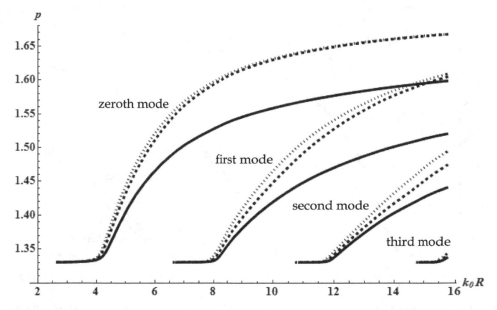

Fig. 9. Dispersion relation p vs $k_0 R$ for the first four modes at intensities $q = 0$ (solid line), $q = 5$ (dashed line) and $q = 10$ (dotted line). The minimum value of vertical axis takes places at $p = n_c = 1.33$.

3.3.3 Electrical control of nonlinear TM modes

In the most general case in which $\lambda \geq 0$ and $q \geq 0$ we are able to tune nonlinear TM modes by varying the uniform electric field represented by E^{dc}. It is expected that, nematic configuration, propagating modes, dispersion relation and cut-off frequencies can be adjusted by modifying the applied electrical field and by modulating the amplitude E_0 of propagating optical field. As said above, while the cut-off frequencies depend directly on λ, ω_c depend inversely on parameter q (see Fig. 2 and Fig. 6). This influence is also observed over the dispersion relation for two different cases: i) the curves acquire larger values of frequencies $k_0 R$ as external electric field λ increases (see Fig 5), whereas ii) the curves adopt smaller values of frequencies as q gets higher (Fig. 9). These two controlling parameters have specific roles on the tuning of the different optical properties of the cylindrical waveguide.

In Fig. 10, we plot the same curves of Fig. 5 but now for nonlinear TM modes for which $q = 10$.

Notice how the influence of applied field over the relation dispersion is modest in comparison to that of Fig. 5 whose curves were clearly modified by the parameter λ. In this particular case, the strength of transverse modes are striving against axial uniform electric field. This result permit us the tuning of optical nonlinear modes in a more precise manner, since the tuning range of λ for changing cut-off frequencies is wider than for the linear mode.

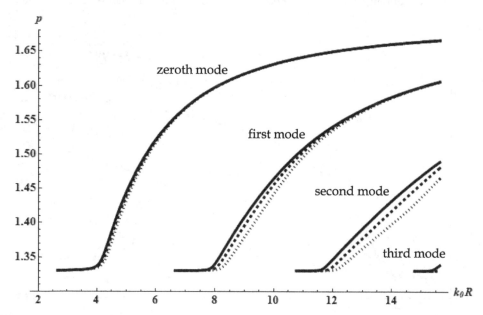

Fig. 10. Dispersion relation p vs $k_0 R$ for the first four modes at intensities $\lambda = 0$ (solid line), $\lambda = 2$ (dashed line) and $\lambda = 4$ (dotted line). The minimum value of vertical axis takes places at $p = n_c = 1.33$.

4. References

Allia P., Oldano C. & Trossi L.(1987). Light propagation in anisotropic stratified media in the quasi adiabatic limit, *Mol. Cryst. Liq. Cryst.* 143, 17-29.

Avendaño C.G.& Reyes J. A.(2004). Spatial Solitons in Chiral Media *Phys. Rev. E* 70, 061701-1-6.

Avendaño C.G. & Reyes J. A. (2006). Wave mixing and spatial structures in cholesteric liquid crystals, *Rev. Mex. Fis. ,S* 52 ,5, 23-31.

Avendaño C.G. & Reyes J.A. (2010). Nonlinear TM Modes in a Cylindrical Liquid Crystal Waveguide, *Opt. Commun.* 283, 24, 5016-5020.

Becchi M., Ponti S., Reyes J. A. & Oldano C. (2004). Defect modes in helical photonic crystals: An analityc approach, *Phys. Rev. B* 70, 033103-7.

Berreman D.W. & Scheffer T.J. (1970). Reflection and transmission by single-domain cholesteric liquid crystal films: theory and verification, *Mol. Cryst. Liq. Cryst.* 11, 395-405.

Born M. & Wolf E.(1975). Principles of Optics, Pergamon, Oxford, UK.

Chen S.-H. & Chen T.-J. (1994). Observation of mode selection in a radially anisotropic cylindrical waveguide with liquid-crystal cladding , *Appl. Phys. Lett.* 64, 1893-1895.

Corella-Madueño A. & Reyes J. A. (2006). Electrically controlled liquid crystal fiber, *Opt. Commun.* 264, 1, 148-155.

Corella-Madueño A., Castellanos-Moreno A., Gutiérrez-López S., Rosas R. A., Reyes J. A. (2008). Threshold field for a nematic liquid cristal confined two coaxial cylinders, *Phys. Rev. E* 78, 022701-4.

Corella-Madueno A. & Reyes J. A. (2008). Hydrodynamically controlled optical propagation in a nematic fiber, *Physica B*, 403 ,1949-1955.

Crawford G. P., Allender D. W. & Doane J. W. (1992). Surface elastic and molecular-anchoring properties of nematic liquid crystals confined to cylindrical cavities, *Phys. Rev. A*, 45, 8693-8708.

de Gennes P.G & Prost J. (1993). The Physics of Liquid Crystals. Clarendon Press, Oxford, UK.

Durbin S. D., Arakelian S.M. & Shen Y. R. (1981). Optical-field-induced birefringence and Freedericksz transition in a nematic liquid crystal, *Phys. Rev. Lett.*, 47, 1411-1414.

Frank F.C. (1958). Liquid crystal. On the theory of liquid crystals, *Faraday Soc. Discuss.* 25, 19-28.

Garcia C., Garza-Hume C., Minzoni A. A., Reyes J. A., Rodriguez R. F. & Smith N. F. (2000). Active propagation and cut-off for low TM modes in a nonlinear nematic waveguide, *Physica D*, 145, 144-157.

Jackson D. (1984). Classical Electrodynamics , Wiley, New York.

Karpierz M.A., Sierakowski M. & Wolinsky T.R. (2002). Light deam propagation in twisted nematics nonlinear waveguides, *Mol. Cryst. Liq. Cryst.*, 375, 313–320.

Khoo I.C. & S. T. Wu S. T.(1993). Optics and nonlinear optics of liquid crystals, World Scientific, Singapore.

Lin H. & Palffy-Muhoray P.(1994). Propagation of TM modes in a nonlinear liquid-crystal waveguide, *Opt. Lett.* 19, 436-438.

Long X.W., Hu W., Zhang T., Guo Q., Lan S. & Gao X.C. (2007). Theoretical investigation of propagation of nonlocal spatial soliton in nematic liquid crystals *Acta Phys. Sin.* 56, 1397-1403

Marcuvitz N. &. Schwigner J.(1951). On the representation of the electric and magnetic fields produced by currents and discontinuities in wave guides I, *J. Appl. Phys.*, 22, 806-819.

McLaughlin D. W., Muraki D. J. & Shelley M. J. (1996). Self-focussed optical structures in a nematic liquid crystal, *Physica D*, 97, 4, 471-497.

Moloney A. C. & Newell J. V. (1992). Nonlinear Optics, Addison Wesley, New York.

Ong H. L. (1987). Wave propagation in layered-inhomogeneous planar anisotropic media: geometrical-optics approximation and its application, *Mol. Cryst. Liq. Cryst.*, 143 , 83-87.

Oseen C. W. (1933). The theory of liquid crystals, *Trans. Faraday Soc.* 29, 883-889.

Reyes J. A. & Rodriguez R. F. (2000). Pulsed Beams in a Nonlinear Nematic Fiber, Physica D, 101, 333-343

Santamato E., G. Abbate G., Maddalena P. & Marrucci L. (1990). Laser-induced nonlinear dynamics in a nematic liquid-crystal film, *Phys. Rev. Lett.*, 64, 1377-1380.

Santamato E. and Y. R. Shen Y.R. (1987) Pseudo-Stokes parameter representation of light propagation in layered inhomogeneous uniaxial media in the geometric optics approximation, *J. Opt. Soc. Am.* 4, 356-359.

Shelton J. W. & Shen Y. R. (1972). Study of Phase-Matched Normal and Umklapp Third-Harmonic- Generation Processes in Cholesteric Liquid Crystals, *Phys. Rev.A*, 5, 1867-1882

Tabiryan N. V., Sukhov A. V. & Zel'dovich B. Y. (1986). Orientational optical nonlinearity of liquid-crystal, *Mol. Cryst.Liq. Cryst.*, 136, 1-139.

Zel'dovich B. Y., Pilipetskii N. F., Sukhov A. V. & Tabiryan N. V. (1980). Giant optical nonlinearity in the mesophase of a nematic liquid crystal, *JETP lett.*, 31, 5, 263-267.

Zwillinger D. (1989). Handbook of Differential Equation, Academic Press, New York

9

Nonlinear Absorption by Porphyrin Supramolecules

Kazuya Ogawa[1] and Yoshiaki Kobuke[2]
Nara Institute of Science and Technology
[1]*Interdisciplinary Graduate School of Medical and Engineering,*
University of Yamanashi,
[2]*Institute of Advanced Energy, Kyoto University,*
Japan

1. Introduction

Two-photon absorption (2PA) is a nonlinear optical process in which two photons are simultaneously absorbed to promote a molecule to the excited state by combination of their energy. 2PA can occur even at wavelengths where one-photon absorption does not take place. Because of quadratic dependence of 2PA on the incident light intensity, the maximum absorption occurs at the focal point of laser allowing high spatial selectivity. These features can find a variety of optical applications such as photodynamic therapy (PDT) (Bhawalkar et al., 1997; E.A. Wachter et al., 1998), 3D optical data storage (Parthenopoulos & Rentzepis, 1989; Strickler & Webb, 1991), and optical limiting (Sutherland, 2003). 2PA was first predicted by Maria Göppert-Mayer in 1931 (Göppert-Mayer, 1931) and was demonstrated experimentally by Kaiser and Garrett using Ruby laser (Kaiser & Garrett, 1961). However, the study on 2PA materials had been inactive until 1990's. After that, new classes of organic molecules exhibiting large 2PA cross section values ($\sigma^{(2)}$) have been reported and the strategies employing donor/acceptor sets with a π-conjugation system in a symmetric (D-π-D or A-π-A) (Albota et al., 1998) or asymmetric (D-π-A) arrangement (Reinhardt et al., 1998) have been proposed.

Porphyrins are attractive target materials for 2PA applications because they have a highly conjugated 18π-electron system leading to a small HOMO-LUMO energy difference. Further, it is interesting in view of visible-light (the Soret band around 400 nm and Q band around 500-700 nm) absorbing and emitting materials as candidates for opto-electronics application as well as nonlinear optics (NLO) including 2PA materials. Novel NLO materials may not be obtained from simple monomeric porphyrins, but be produced when strong electronic interactions between porphyrins are induced by self-assembly to bring porphyrins close together or connecting their π-conjugation systems.

A lot of multiporphyrin systems have been reported either by covalent or noncovalent approaches (Chambron et al., 1999; Chou et al., 1999; Ogawa & Kobuke, 2004) and some of them showed such the strong electronic interactions between porphyrins. In most of covalent approaches, strong interactions have been found in linear porphyrin arrays by connecting porphyrins at *meso*-positions (Osuka & Shimidzu, 1997; Aratani et al., 2005; Tsuda & Osuka, 2001; Anderson, 1994; Anderson, 1999; Lin et al., 1994).

In contrast to the covalent approaches, noncovalent approaches allow easy construction of multiporphyrin arrays. However, studies on self-assembled porphyrin arrays exhibiting strong excitonic and electronic interactions are limited because it is hard to arrange porphyrins in an appropriate position to invoke strong excitonic interaction between porphyrins.

We have reported that zinc imidazolylporphyrin **1** (Fig. 1) allows formation of stable slipped cofacial dimer **2D** through complementary coordination of the imidazolyl to zinc in another porphyrin with a stability constant over 10^{11} M^{-1} (Kobuke & Miyaji, 1994). Then, we challenged to construct one-dimensional supramolecular linear porphyrin arrays by connecting the complementary coordination dimer units in a linear fashion. As a result, a giant supramolecular porphyrin array **3P** of over 800 porphyrin units could be obtained by linking two imidazolylporphyrin units directly at the *meso* positions (Ogawa & Kobuke, 2000). The polymeric structure can easily be cleaved by adding coordinating solvents such as MeOH or pyridine, and reorganized again by removing the solvents. When this reorganization was performed in the presence of **3P** and another imidazolylporphyrinatozinc dimer **4D**, oligomer **5n** having terminal units of **4M** was obtained. Thus, the reorganization can serve an efficient method for introducing appropriate donor and/or acceptor groups at the molecular terminals of the array as a substituent R$_2$. For example, when freebase porphyrin was used as an acceptor, large enhancements of the real part of the molecular second hyperpolarizability were observed (Ogawa et al., 2002). However, almost no nonlinear absorption was observed in femtosecond optical Kerr effect (OKE) measurements at the off-resonant wavelength of 800 nm.

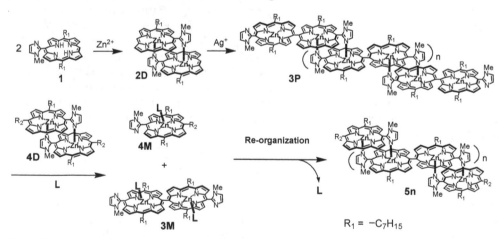

Fig. 1. Formation of self-assembled porphyrin dimer and giant porphyrin array by complementary coordination of imidazolyl to zinc.

2. Strong 2PA of conjugated porphyrins by self-coordination

In order to extend the use of this supramolecular porphyrin system to 2PA materials (i.e. to obtain a large 2PA cross section) we designed a novel porphyrin assembly **7D** (Fig. 2)

(Ogawa et al., 2003) according to the general strategy as mentioned above. At first, two porphyrins were connected by butadiynylene to allow π-conjugation by taking a coplanar orientation of two porphyrins. In the case of *meso-meso* linked bisporphyrins shown in Fig. 1, orthogonal orientation between porphyrins prevents the desired porphyrin-porphyrin π-conjugation. The conjugated porphyrin arrays, covalently linked by butadiynylene and ethynylene linkages (Piet et al., 1997; Anderson, 1994; Anderson, 1999; Thorne et al., 1999; Screen et al., 2002; Karotki et al., 2004; Drobizhev et al., 2004; Lin et al., 1994; Lin et al., 1995; Angiolilloet al., 2004), are assumed to be converted to cumulenic structures upon photoexcitation. This may make the absorption of the second photon favorable and enhance their nonlinear optical properties. Next, free base porphyrins as electron acceptors were introduced at both terminals of the array to induce molecular polarization.

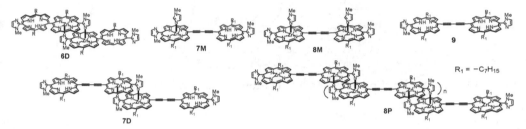

Fig. 2. Structures of compounds investigated.

2.1 Effect of the expansion of π -conjugation between porphyrins on 2PA

The butadiynylene linkage in **7D** allows a coplanar orientation between bisporphyrins, leading to the expansion of π-conjugation. On the other hand, in the case of **6D**, two porphyrins directly connected at *meso*-positions are almost orthogonal to each other, preventing π-conjugation between two porphyrins. Fig. 3 shows one-photon absorption spectra of **6D** (dotted line) and **7D** (bold solid line) in CHCl₃. Both self-assemblies show strong one-photon absorption in the range of 400 nm to 500 nm, which corresponds to the S2 state (the Soret band). On the other hand, only very weak absorptions are observed over 800 nm. These observations suggest that two photons at the wavelength over 800 nm will be absorbed simultaneously to promote the molecules to the S2 state. The Q-band (the S1 state) of **7D** was red-shifted to 740 nm compared to that of **6D** (660 nm), and was also significantly intensified suggesting the expansion of the porphryin-porphyrin π-conjugation due to the butadiynylene linkage.

The 2PA cross sections were measured using an open aperture Z-scan method (Sheik-Bahae et al., 1990) at wavelengths from 810 to 1300 nm, with a femtosecond optical parametric amplifier. This technique detects nonlinear absorption, i.e. two- or multi-photon absorption, by scanning the sample around the focal point along the direction of laser beam (Z-axis). The nonlinear absorption is observed most strongly at the focal point, where the peak intensity of the incident light becomes maximum. The $\sigma^{(2)}$ value can be estimated by curve fitting using theoretical equations.

2PA spectra of **6D** and **7D** measured in CHCl₃ are shown in Fig. 4. The maximum $\sigma^{(2)}$ values were obtained as 370 GM at 964 nm for **6D** and 7,600 GM at 887 nm for **7D**, respectively. The maximum value obtained for **7D** is almost 20 times larger than that for **6D**, showing that the expansion of π-conjugation between porphyrins by introducing the butadiynylene linkage is the most important factor to enhance the $\sigma^{(2)}$ value. The value of 7,600 GM was the largest class among reported organic compounds measured in femtosecond time scale at that time.

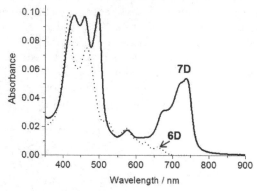

Fig. 3. One-photon absorption spectra of **6D** (dotted line) and **7D** (bold solid line) in CHCl₃.

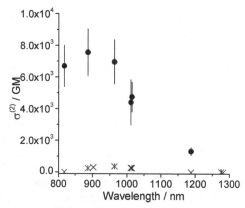

Fig. 4. 2PA spectra of **6D** (cross) and **7D** (circle) measured using femtosecond pulses in CHCl₃.

2.2 Effects of complementary coordination and monozinc metalation on the enhancement of the 2PA cross section

As described in the above section, the expansion of π-conjugation between two porphyrins is one of the significant reasons for the enhancement of the $\sigma^{(2)}$ value. From a viewpoint of the relationship between structure and 2PA property, it is interesting to examine the effect of the complementary coordination of imidazolyl to zinc and the monozinc metalation, which induces the molecular polarization. Therefore, we examined the 2PA properties of bisporphyrins in the forms of monozinc complex **7M**, di-zinc complex **8M**, and free base **9** in CHCl₃ solution containing 3,000 equivalents of 1-methylimidazole, which acted as a competing ligand to lead the cleavage of the complementary coordination. The 2PA absorption spectra of **7M**, **8M**, and **9** are shown in Fig. 5. These compounds without the complementary coordination also exhibited relatively large $\sigma^{(2)}$ values of 1,800, 1,200, and 1,000 GM, respectively. The $\sigma^{(2)}$ value of **7M** was almost twice of those obtained for **8M** and **9**, suggesting that monozinc metalation is effective for 2PA enhancement. Since free base porphyrin works as an electron acceptor against zinc porphyrin, the monozinc metalation induces the molecular polarization which may cause intramolecular charge transfer in the 2PA transition process. On the other hand, the $\sigma^{(2)}$ value of 1,800 GM obtained for **7M** was four times smaller than that for **7D**. This indicates that the complementary coordination is another effective enhancement factor. In **7D**, the complementary coordination brings in the extended electronic communication through the coordination bonds. Further, the complementary coordination may also contribute to the larger $\sigma^{(2)}$ value of **7D** by the enhancement of the transition dipole moments as a result of the excitonic coupling (McRae & Kasha, 1958).

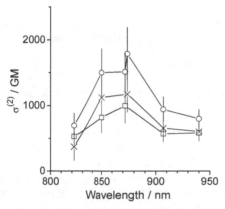

Fig. 5. 2PA spectra of **7M** (circle), **8M** (cross), and **9** (square) in CHCl₃ with 1-methylimidazole (1.5 M).

2.3 Effect of polymerization on the enhancement of the 2PA cross section

Reaction of free base **9** with excess amount of zinc acetate give di-zinc complex, which allows the formation of one-dimensionally propagated polymer **8P** by the successive complementary coordination of imidazolyl to zinc. Molecular weight of **8P** was analyzed by using gel permeation chromatography (GPC) with a column having an exclusion limit of 500,000 Dalton eluted by CHCl₃. The mean molecular weights, M_w and M_n of **8P** were estimated by comparing the data with those of polystyrene standards as 200,000 and 150,000, respectively.

The 2PA cross section of **8P** was measured in CHCl₃. Strong two-photon absorption was also observed in the Z-scan experiment and the $\sigma^{(2)}$ value per biszinc unit was estimated to be 4,000 GM at 873 nm. The enhancement factor for polymerization is 3.3 compared with the value obtained for **8M**. The mean $\sigma^{(2)}$ value of polymer **8P** was estimated to be ~400,000 GM, using the M_n value of 150,000 corresponding to 110 bisporphyrin units. The two-photon absorption spectra of **8P** are shown in Fig. 6. This value is extremely large compared with the value for **8M** with an enhancement factor of 360. Although the $\sigma^{(2)}$ value per dimer unit of **8P** shows the enhancement factor of 3.3 compared with that of **8M**, the value of 4,000 GM is almost the same as that of **7D** (3,800 GM), suggesting that the elongation effect more than three dimer units is not so significant for the 2PA enhancement.

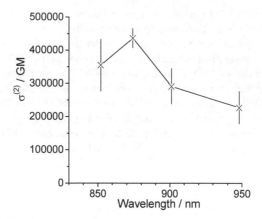

Fig. 6. 2PA spectra of **8P** in CHCl₃.

2.4 Nonlinear absorption measured with nanosecond pulses

Generally, the $\sigma^{(2)}$ value obtained with nanosecond laser pulses is 2 or 3 orders of magnitude larger than that measured using femtosecond pulses due to the excited state absorption (ESA) (Swiatkiewicz et al., 1998; Kim et al., 2000; Lei et al., 2001). This contribution is difficult to remove from the Z-scan data. However, on considering application of 2PA materials such as 2PA-PDT or 3-D optical memory, the nanosecond pulse system may be more appropriate because of availability and the easiness of operation. Therefore, we examined nonlinear absorption properties of nanosecond region for **7D** and **8P** by Z-scan measurements with a nanosecond optical parametric oscillator (OPO) pumped by a Nd:YAG laser. Fig. 7 shows plots of two-photon absorbance q_0 (Ogawa et al., 2005) against incident light intensity I_0 for **8P** at 850 nm. The pulse energy was varied between 1.0 mJ and 2.1 mJ. The value of q_0 was almost proportional up to I_0 of 1.1×10^{14} W/m², corresponding to

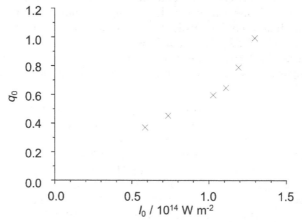

Fig. 7. Plots of the two-photon absorbance q_0 against peak intensity I_0 for **8P** at 850 nm with nanosecond pulses.

I_0 = 15 mW. This result shows that the ESA is significant at the high intensity region over 15 mW. Therefore, the effective $\sigma^{(2)}$ ($^{eff}\sigma^{(2)}$) spectrum was measured at the lowest intensity of 1.0 mJ as shown in Fig. 8. These spectra were similar to those obtained using femtosecond measurements and absorption maxima were observed at 890 nm in both compounds. The maximum $^{eff}\sigma^{(2)}$ values for **7D** and **8P** were estimated as 210,000 and ~22,000,000, respectively, which were 30 to 50 times larger than those observed by femtosecond pulses, indicating the presence of the ESA contribution.

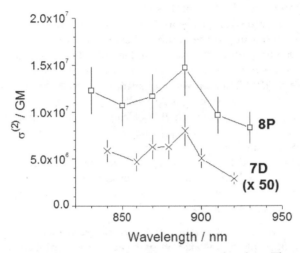

Fig. 8. 2PA spectra of **7D** and **8P** measured with nanosecond pulses. The pulse energy was 1.0 mJ for all the measurements.

2.5 Three-photon absorption

The higher-order nonlinear absorption than 2PA was found for **7D** at wavelengths longer than 1180 nm. The most probable higher-order nonlinear absorption process observed on using the femtosecond pulses originates from three-photon absorption (3PA). Furthermore, the wavelengths were almost three times long as that of the Soret band, suggesting 3PA. 2PA and 3PA can occur simultaneously in this wavelength range. When the observed data were analyzed as only 3PA was taken into account, a linear relationship between 3PA parameter p_0 (Tykwinski et al., 2002; Ogawa et al., 2005) and I_0 was obtained. This linear relationship does not necessarily mean the observation of the intrinsic 3PA (i.e. the simultaneous absorption of three photons), since two-step 3PA (e.g. 2PA followed by an excited state absorption process) also shows the same linear relationship. Although the possibility of two-step 3PA process cannot be completely ruled out, the observation may originate from the intrinsic 3PA process because the $\sigma^{(2)}$ values decreased as the irradiation wavelength approached the wavelength range in which 3PA was observed, and furthermore this range corresponds to three times the wavelength of the Soret band. The 3PA cross sections were estimated to be 7.1×10^{-77} m^6s^2 and 1.8×10^{-77} m^6s^2 at 1190 and 1280 nm, respectively. Materials exhibiting 3PA may be interesting for further high-resolution 3D optical memory, 3D fabrication, and PDT applications with merit of using longer NIR wavelength light.

2.6 Effects of water-solubilization and ethynylene connection on 2PA

Photodynamic therapy (PDT) is a medical treatment of cancers, which uses a photosensitizer without surgery, and is a one of the possible 2PA applications as described in Introduction. The penetration depth of visible light used in currently available PDT (630 nm) is limited only to reach tissue surface due to absorption and scattering by biological tissue, indicating that this method cannot be applied to the treatment of deep cancers. However, the penetration depth can be improved by using longer wavelength range of 700-1500 nm which is relatively transparent for biological tissue and called as the optical window. Since porphyrin compounds have strong one-photon absorption bands between 400 and 500 nm (the Soret band) corresponding to the combined energy of two photons in the wavelength range from 800 to 1000 nm, which is just laid in the optical window, 2PA using porphyrins is suitable for PDT for deep cancers. Quadratic dependence on the laser intensity is another advantage of the use of 2PA. This allows high spatial selectivity by focusing the laser beam at the target point and prevents damages to healthy tissue. So, we have studied 2PA-PDT using conjugated porphyrins having high 2PA efficiency. First, compound **7D** was modified to be solubilized in water. Because porphyrins are tend to stack in water and this may affect optical properties of compounds, effect of water-solubilization on 2PA was examined. The water-solubility was obtained by introducing a carboxylic group instead of heptyl at each *meso*-position (**10D** (Fig. 9)) (Ogawa et al., 2006). Further, ethynylene-linked compound **11D**, obtained from a direct hetero-coupling reaction between donor zincporphyrin and acceptor freebase porphyrin, was also synthesized.

The $\sigma^{(2)}$ values in water were measured by a femtosecond open aperture Z-scan method at 850 nm. Strong nonlinear absorption was also observed at the focal point even in a dilute solution of 0.37 mM. The $\sigma^{(2)}$ value of **10D** was determined as 7,500 GM. This value is almost same as that of **7D** in chloroform, indicating no water-solubilization effect on 2PA. The $\sigma^{(2)}$ value of ethynylene compound **11D** was also measured as 7,900 GM, being almost equivalent to that of **10D**. These results show that the water-soluble porphyrin assemblies **10D** and **11D**, exhibiting strong two-photon absorption, are possible candidates for the 2PA-PDT agent.

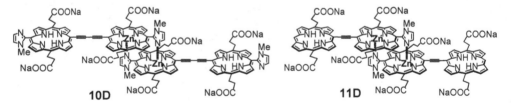

Fig. 9. Structures of water-soluble self-assembled porphyrins **10D** and **11D**.

2.7 Effect of elongation of conjugated bisporphyrin on 2PA

Bis(imidazolylporphyrin) generates not only polymers but also discrete shorter porphyrin arrays by the addition of monomeric imidazolylporphyrins as terminators through the complementary coordination of imidazolyl to zinc (Ogawa & Kobuke, 2000; Kobuke & Ogawa, 2003; Ogawa & Kobuke, 2006). The conjugated bisporphyrins also gives regulated porphyrin arrays by the self-assembly. Consequently, we synthesized a butadiynylene-bridged bisporphyrin monomer **12** and dimer **13** terminated with monomeric porphyrins on both ends and examined the effect of incremental elongation of butadiynylene-linked porphyrin arrays on 2PA (Dy et al., 2008).

Absorption spectra are illustrated in Fig. 11. Zinc-imidazolyl coordinated dimer **14** shows characteristic splitting of the Soret band at 415 and 438 nm. On the other hand, the Soret band of **12** having one butadiynylene-linked bisporphyrin unit exhibits broader and red-shifted peaks at 434, 461, and 495 nm due to not only larger head-to-tail interaction between chromophores but also the expansion of π-conjugation. Furthermore, strong Q-bands corresponding to HOMO-LUMO absorption were observed at longer wavelengths of 666 and 728 nm, also indicating the expansion of π-conjugation. Q-bands of **13** which consists of two butadiynylene-linked bisporphyrin units appeared at 670 and 733 nm which are slightly red-shifted and amplified compared to **12**. The amplification of Q-bands is not two times but almost three times the intensity, indicating stronger dipole moments caused by excitonic interaction between the two bisporphyrins, and may contribute to larger resonance enhancement for **13**.

The 2PA spectra were measured using the open-aperture z-scan method with 120 fs pulses at off-resonant wavelengths from 820 to 940 nm, as described above. Compounds **12** and **13** exhibited maximum $\sigma^{(2)}$ values of 10,000 and 61,000 GM at 870 nm, respectively (Fig. 12), showing a six times enhancement of $\sigma^{(2)}$. Self-assembled porphyrin dimer **14** was not measured due to its too weak 2PA (less than 20 GM). The large $\sigma^{(2)}$ value observed for **12** is mainly due to the expansion of π-conjugation by the butadiynylene linkage as evidenced by the large and red-shifted Q-band (Fig. 11).

The insertion of another butadiynylene-linked bisporphyrin unit by complementary coordination resulted in a further enhanced $\sigma^{(2)}$ value. As described in the one-photon absorption spectra, the amplification of the Q-band of **13** is almost three times the value of **12** due to the larger transition dipole moments caused by the excitonic interaction between two butadiynylene-linked bisporphyrins. Since the $\sigma^{(2)}$ value is proportional to the square of the transition dipole moment of the one-photon absorption (Birge & Pierce, 1979), the amplification of the Q-band may contribute to larger enhancement for **13**.

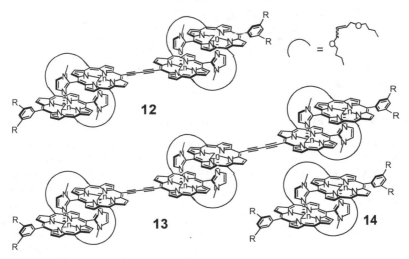

Fig. 10. Structures of compounds **12**, **13**, and **14**.

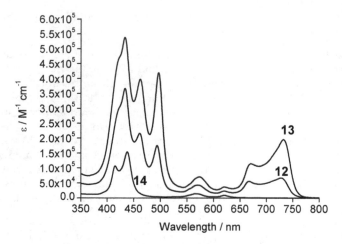

Fig. 11. One-photon absorption spectra of compounds **12**, **13**, and **14**.

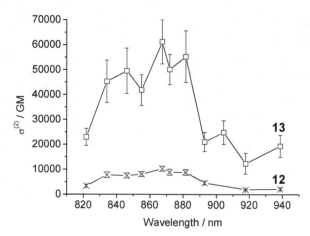

Fig. 12. 2PA spectra of **12** and **13**.

2.8 Nonlinear absorption by self-assembled porphyrin-phthalocyanine conjugates

As discussed above, we have developed supramolecular conjugated porphyrins linked with a ethynylene or butadiynylene bond exhibiting the largest class of $\sigma^{(2)}$ values reaching an order of 10^4 GM and relatively strong 3PA, too. Then, we have developed further multichromophore systems, in which porphyrin and other chromophores are connected using an ethynylene inducing a molecular polarity. Phthalocyanines are of attractive interest because they have an 18π conjugation system as large as porphyrins and are chemically stable. In this respect, it is interesting to construct a porphyrin-phthalocyanine conjugate using the ethynylene bridge. Here, we introduce large multiphoton absorption, 2PA and 3PA, behavior of self-organized dimers of imidazolylporphyrin-zincphthalocyanine **15**

(H_2(ImPor)-Zn(Pc)) and **16** (Zn(ImPor)-Zn(Pc)) in the antiparallel fashion (Fig 13) (Morisue and Kobuke, 2008; Morisue et al., 2010). A high association constant of complementary coordination of imidazolyl to zinc ($\sim10^{14}$ M^{-1}) allows organization of M(ImPor)-Zn(Pc) into the dimer. The dimer composed of porphyrin and phthalocyanine tetrad gives following advantages for exhibiting strong multiphoton absorption. One is the coplanar π-conjugation between porphyrin and phthalocyanine through the ethynylene bond. The other is the possibility of the resonance enhancement of 2PA transition, because the Q band of phthalocyanine would give a small detuning energy for a 2PA transition in the near-IR range. Furthermore, the dimer may form a 2D quadrupolar structure different from the 1D ones, for example D-π-A-π-D or A-π-D-π-A. In addition, electron-withdrawing nature of porphyrin can be tuned further by metalation of the porphyrin center, for example, **15** having free base porphyrin gives higher charge-transfer ability than **15** having zincporphyrin.

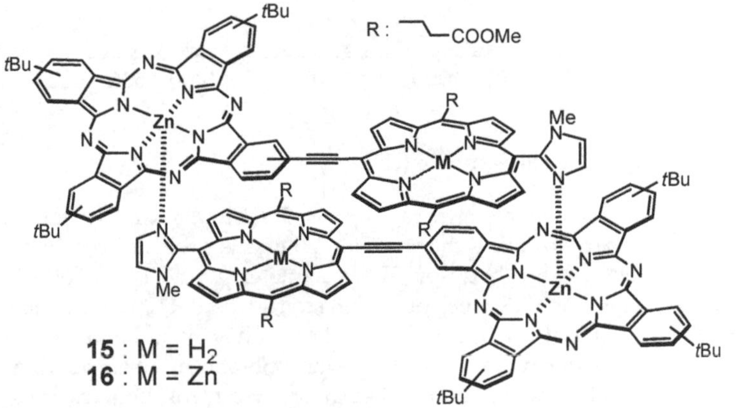

Fig. 13. Structures of the antiparallel dimers **15** (M=H_2) and **16** (M=Zn). The dimer is a regioisomeric mixture due to the presence of the regioisomeric mixtures of ethynylene linker and *tert*-butyl groups at the β-position of phthalocyanine.

Nonlinear absorption (NLA) was measured with the same femtosecond Z-scan system as above Chapters varying the incident laser wavelength from 800 to 1550 nm. The NLA behavior depended on the wavelength range. At a wavelength shorter than 996 nm, the simultaneous 2PA process was dominant. In the wavelength region shorter than 900 nm where the edge of one-photon absorption remains, the saturable absorption (SA) of one-photon absorption was observed. The Z-scan data was analyzed by considering the SA process (Morisue et al., 2010). As shown in 2PA spectra (Fig. 14), $\sigma^{(2)}$ value increased with decreasing the incident wavelength and no peak top was observed. The maximum $\sigma^{(2)}$ values in the wavelength region where SA and 3PA are negligible (900–996 nm) were 16,000 and 7,000 GM at 900 nm for **15** and **16**, respectively. These $\sigma^{(2)}$ values are relatively large compared to those of related compounds as discussed in above Chapters. The $\sigma^{(2)}$ value of **15** having free base porphyrin is twice larger than that of **16** having zincporphyrin, suggesting that larger molecular polarity amplified the $\sigma^{(2)}$ value for **15**. The obtained large values suggest considerable π-delocalization over tetrad through complementary coordination between imidazolylporphyrin-zincphthalocyanine dimers. Unfortunately, the comparable value of the dissociated dimer could not be obtained due to its extremely large association constant of 10^{14} M^{-1}.

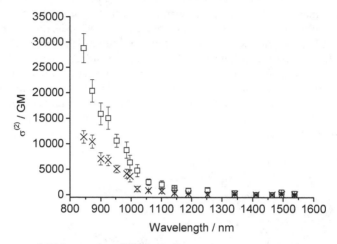

Fig. 14. 2PA spectra of **15** (square) and **16** (cross).

On the other hand, in the wavelength region longer than 996 nm, the higher-order NLA other than 2PA was observed for all the samples. When the observed data were analyzed as 3PA, a good linear relationship between 3PA parameter p_0 and I_0 was obtained. The 3PA spectra are shown in Fig 15. 3PA were observed around 1000–1100 nm, which corresponds to three times the wavelength of the Soret band of phthalocyanine part (around 350 nm). Since 3PA has the same parity selection rule with one-photon absorption, a one-photon-allowed excited state is always 3PA-allowed even for the centrosymmetric system unlike the case of 2PA. Therefore, the final state of the 3PA transition is considered to be the Soret band of phthalocyanine part. The maximum $\sigma^{(3)}$ values were estimated as 7.6×10^{-76} cm^6 s^2 at 1020

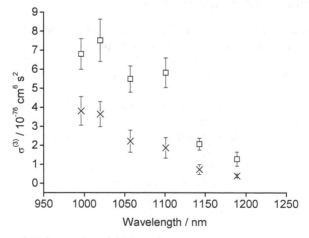

Fig. 15. 3PA spectra of **15** (square) and **16** (cross).

nm and 3.8×10^{-76} cm^6 s^2 at 996 nm for **15** and **16**, respectively. The values are almost 10 times larger than those obtained for **7D**. This large $\sigma^{(3)}$ values may arise from the resonance enhancement of the 2PA state or the two-step 3PA process: i.e., the 2PA followed by the excited state absorption (ESA).

3. Conclusion

Recent progress in the nonlinear absorption (2PA and 3PA) properties of self-assembled porphyrin arrays has been reviewed. These studies demonstrate that self-coordination for constructing porphyrin arrays based on complementary coordination of imidazolyl to zinc is a convenient and powerful tool enhancing the nonlinear absorption. We believe that these investigations open the way for further exploration of new types of nonlinear absorption materials. Moreover, the enhancement of the cross-sections will make possible the production of future 2PA applications.

4. Acknowledgment

The authors acknowledge Drs. K. Kamada and K. Ohta at Photonics Research Institute, National Institute of Advanced Industrial Science and Technology for femtosecond Z-scan measurements and discussions, and all the group members of our laboratory at Nara Institute of Science and Technology.

5. References

Albota, M.; Beljonne, D.; Brédas, J.-L.; Ehrlich, J. E.; Fu, J.-Y.; Heikal, A. A.; Hess, S. E.; Kogej, T.; Levin, M. D.; Marder, S. R.; McCord-Maughon, D.; Perry, J.W.; Röckel, H.; Rumi, M.; Subramaniam, G.; Webb, W. W.; Wu, X.-L. & Xu, C. (1998). Design of Organic Molecules with Large Two-Photon Absorption Cross Sections, *Science*, 281, 1653-1656.

Anderson, H. L. (1999). Building molecular wires from the colours of life: conjugated porphyrin oligomers, *Chem. Commun.* 2323-2330.

Angiolillo, P. J.; Uyeda, H. T.; Duncan, T. V. & Therien, M. J. (2004). Impact of electronic asymmetry on photoexcited triplet-State spin distributions in conjugated porphyrin oligomers probed via EPR spectroscopy, *J. Phys. Chem. B*, 108, 11893-11903.

Anderson, H. L. (1994). Conjugated Porphyrin Ladders, *Inorg. Chem.* 33, 972-981.

Aratani,N.; Takagi, A.; Yanagawa, Y.; Matsumoto, T.; Kawai, T.; Yoon, Z. S.; Kim, D.; Osuka, A. (2005). Giant meso–meso-linked porphyrin arrays of micrometer molecular length and their fabrication, *Chem. Eur. J.*, 11, 3389-3404.

Bhawalkar, J. D.; Kumar, N. D.; Zhao,C. F.; Prasad, P. N. (1997). Two-photon photodynamic therapy. *J. Clin. Laser Med. Surg.* 15, 201-204.

Birge, R. R. & Pierce, B. M. (1979). A theoretical analysis of the two-photon properties of linear polyenes and the visual chromophores, *J. Chem. Phys.*, 70, 165-178.

Chambron,J.-C.; Heitz, V. & Sauvage, J.-P. (1999). Noncovalent Multiporphyrin Assemblies, In: *The Porphyrin Handbook*, Kadish,K.; Smith, K. M. & Guilard, R. (Eds), Vol. 6, 1-42, Academic Press, New York.

Chou, J.-H.; Kosal, M. E.; Nalwa, H. S.; Rakow, N. A. & Suslick, K. S. (1999). Applications of Porphyrins and Metalloporphyrins to Materials Chemistry in: *The Porphyrin Handbook*, Kadish,K.; Smith, K. M. & Guilard, R. (Eds), Vol. 6, Academic Press, New York, 43-132.

Drobizhev, M.; Stepanenko, Y.; Dzenis, Y.; Karotki, A.; Rebane, A.; Taylor, P. N. & Anderson, H. L. (2004). Understanding strong two-photon absorption in π-conjugated porphyrin dimers via double-resonance enhancement in a three-level model, *J. Am. Chem. Soc.*, 126, 15352-15353.

Dy, J. T.; Ogawa, K.; Kamada, K.; Ohta, K. & Kobuke, Y. (2008). Stepwise elongation effect on the two-photon absorption of self-assembled butadiyne porphyrins, *Chem. Commun.*, 3411-3413.

Göppert-Mayer, M. (1931). Über Elementarakte mit zwei Quantensprüngen, *Ann. Phys. 9*, 273-294.

Kaiser, W. & Garrett, C. G. B. (1961). Two-photon excitation in $CaF_2:Eu^{2+}$, *Phys. Rev. Lett. 7*, 229-231.

Karotki, A.; Drobizhev, M.; Dzenis, Y.; Taylor, P. N.; Anderson, H. L. & Rebane, A. (2004). Dramatic enhancement of intrinsic two-photon absorption in a conjugated porphyrin dimer, *Phys. Chem. Chem. Phys.*, 6, 7-10.

Kim, O.-K.; Lee, K.-S.; Woo, H. Y.; Kim, K.-S.; He, G. S.; Swiatkiewicz, J. & Prasad, P. N. (2000). New class of two-photon-absorbing chromophores based on dithienothiophene, *Chem. Mater.*, 12, 284-286.

Kobuke, Y. & Miyaji, H. (1994). Supramolecular organization of imidazolyl-porphyrin to a slipped cofacial dimer, *J. Am. Chem. Soc.*, 116, 4111-4112.

Kobuke, Y. & Ogawa, K. (2003). Porphyrin supramolecules for artificial photosynthesis and molecular photonic/electronic materials, *Bull. Chem. Soc. Jpn.*, 76, 689-708.

Lei, H.; Wang, H. Z.; Wei, Z. C.; Tang, X. J.; Wu, L. Z.; Tung, C. H. & Zhou, G. Y. (2001). Photophysical properties and TPA optical limiting of two new organic compounds, *Chem. Phys. Lett.*, 333, 387-390.

Lin, V. S.-Y.; DiMagno, S. G. & Therien, M. J. (1994). Highly conjugated, acetylenyl bridged porphyrins: new models for light-harvesting antenna systems, *Science*, 264, 1105-1111.

Lin, V. S.-Y. & Therien, M. J. (1995). The role of porphyrin-to-porphyrin linkage topology in the extensive modulation of the absorptive and emissive properties of a series of ethynyl- and butadiynyl-rridged bis- and tris(porphinato)zinc chromophores, *Chem. Eur. J.*, 1, 645-651.

McRae, E. G. & Kasha, M. (1958). Enhancement of phosphorescence ability upon aggregation of dye molecules, *J. Chem. Phys.*, 28, 721-722.

Morisue, M. & Kobuke, Y. (2008). Tandem cofacial stacks of porphyrin–phthalocyanine dyads through complementary coordination, *Chem. Eur. J.*, 14, 4993-5000.

Morisue, M.; Ogawa, K.; Kamada, K.; Ohta, K. & Kobuke, Y. (2010)., *Chem. Commun.*, 2121-2123

Ogawa, K.; Hasegawa, H.; Inaba, Y.; Kobuke, Y.; Inouye, H.; Kanemitsu, Y.; Kohno, E.; Hirano, T.; Ogura, S. & Okura, I. (2006). Water-soluble bis(imidazolylporphyrin) self-assemblies with large two-photon absorption cross sections as agents for photodynamic therapy, *J. Med. Chem.*, 49, 2276-2283.

Ogawa, K. & Kobuke, Y. (2000). Formation of a giant supramolecular porphyrin array by self-coordination, *Angew. Chem. Int. Ed. Engl.*, 39, 4070-4073.

Ogawa,K. & Kobuke, Y. (2004). Self-Assembled Porphyrin Array, In: H.S. Nalwa (Ed), *Encyclopedia of Nanoscience and Nanotechnology*, Vol. 9, American Scientific Publishers, Stevenson Ranch/California, 561-591.

Ogawa, K.; Ohashi, A.; Kobuke, Y.; Kamada, K. & Ohta, K. (2003). Strong Two-Photon Absorption of Self-Assembled Butadiyne-Linked Bisporphyrin, *J. Am. Chem. Soc.*, 125, 13356-13357.

Ogawa,K. & Kobuke, Y. (2006). Construction and photophysical properties of self-assembled linear porphyrin arrays, *J. Photochem. Photobiol. C*, 7, 1-16.

Ogawa, K.; Ohashi, A.; Kobuke, Y.; Kamada, K. & Ohta, K. (2005). Two-photon absorption properties of self-assemblies of butadiyne-linked bis(imidazolylporphyrin), *J. Phys. Chem. B*, 109, 22003-22012.

Ogawa, K.; Zhang, T.; Yoshihara, K. & Kobuke, Y. (2002). Large third-order optical nonlinearity of self-assembled porphyrin oligomers, *J. Am. Chem. Soc.*, 124, 22-23.

Osuka, A. & Shimidzu, H. (1997). *meso, meso*-Linked Porphyrin Arrays, *Angew. Chem. Int. Ed. Engl.* 36, 135-137.

Parthenopoulos, D. A. & Rentzepis, P. M. (1989). Three-Dimensional Optical Storage Memory. *Science*, 245 , 843-845.

Piet, J. J.; Warman, J. M. & Anderson, H. L. (1997). Photo-induced charge separation on conjugated porphyrin chains, *Chem. Phys. Lett.*, 266, 70-74.

Reinhardt, B. A.; Brott, L. L.; Clarson, S. J.; Dillard, A. G.; Bhatt, J. C.; Kannan, R.; Yuan, L.; He, G. S. & Prasad, P. N. (1998). Highly Active Two-Photon Dyes: Design, Synthesis, and Characterization toward Application. *Chem. Mater.*, 10, 1863-1874.

Screen, T. E. O.; Thorne, J. R. G.; Denning, R. G.; Bucknall, D. G. Anderson, H. L. (2002). Amplified Optical Nonlinearity in a Self-assembled double-strand conjugated porphyrin polymer ladder, *J. Am. Chem. Soc.*, 124, 9712-9713.

Sheik-Bahae, M.; Said, A. A.; Wei, T.-H.; Hagan, D. G. & van Stryland, E. W. (1990). Sensitive measurement of optical nonlinearities using a single beam, *IEEE J. Quant. Electr.*, 26, 760-769.

Strickler, J. H. & Webb, W. W. (1991). Three-dimensional optical data storage in refractive media by two-photon point excitation, *Opt. Lett.*, 16, 1780-1782.

Sutherland, R. L. (2003). *Handbook of Nonlinear Optics* (Second edition), Marcel Dekker, New York.

Swiatkiewicz, J.; Prasad, P. N. & Reinhardt, B. A. (1998). Probing two-photon excitation dynamics using ultrafast laser pulses, *Opt. Comm.*, 157, 135-138.

Thorne, J. R. G.; Kuebler, S. M.; Denning, R. G.; Blake, I. M.; Taylor, P. N. & Anderson, H. L. (1999). Degenerate four-wave mixing studies of butadiyne-linked conjugated porphyrin oligomers, *Chem. Phys.*, 248, 181-193.

Tsuda, A. & Osuka, A. (2001). Fully conjugated porphyrin tapes with electronic absorption bands that reach into infrared, *Science*, 293, 79-82.

Tykwinski, R. R.; Kamada, K.; Bykowski, D.; Hegmann, F. A. & Hinkle, R. J. (2002). Nonlinear optical properties of thienyl and bithienyl iodonium salts as measured by the Z-scan technique, *J. Opt. A: Pure Appl. Opt.*, 4, S202-S206.

Wachter, E. A.; Partridge, W. P.; Fisher, W. G.; Dees, H. C.; Petersen, M. G. (1998). Simultaneous two-photon excitation of photodynamic therapy agents, *Proc. SPIE-Into. Soc. Opt. Eng.*, 3269, 68-75.

Permissions

The contributors of this book come from diverse backgrounds, making this book a truly international effort. This book will bring forth new frontiers with its revolutionizing research information and detailed analysis of the nascent developments around the world.

We would like to thank N. V. Kamanina, Dr.Sci., PhD, for lending her expertise to make the book truly unique. She has played a crucial role in the development of this book. Without her invaluable contribution this book wouldn't have been possible. She has made vital efforts to compile up to date information on the varied aspects of this subject to make this book a valuable addition to the collection of many professionals and students.

This book was conceptualized with the vision of imparting up-to-date information and advanced data in this field. To ensure the same, a matchless editorial board was set up. Every individual on the board went through rigorous rounds of assessment to prove their worth. After which they invested a large part of their time researching and compiling the most relevant data for our readers. Conferences and sessions were held from time to time between the editorial board and the contributing authors to present the data in the most comprehensible form. The editorial team has worked tirelessly to provide valuable and valid information to help people across the globe.

Every chapter published in this book has been scrutinized by our experts. Their significance has been extensively debated. The topics covered herein carry significant findings which will fuel the growth of the discipline. They may even be implemented as practical applications or may be referred to as a beginning point for another development. Chapters in this book were first published by InTech; hereby published with permission under the Creative Commons Attribution License or equivalent.

The editorial board has been involved in producing this book since its inception. They have spent rigorous hours researching and exploring the diverse topics which have resulted in the successful publishing of this book. They have passed on their knowledge of decades through this book. To expedite this challenging task, the publisher supported the team at every step. A small team of assistant editors was also appointed to further simplify the editing procedure and attain best results for the readers.

Our editorial team has been hand-picked from every corner of the world. Their multi-ethnicity adds dynamic inputs to the discussions which result in innovative outcomes. These outcomes are then further discussed with the researchers and contributors who give their valuable feedback and opinion regarding the same. The feedback is then collaborated with the researches and they are edited in a comprehensive manner to aid the understanding of the subject.

Apart from the editorial board, the designing team has also invested a significant amount of their time in understanding the subject and creating the most relevant covers. They scrutinized every image to scout for the most suitable representation of the subject and create an appropriate cover for the book.

The publishing team has been involved in this book since its early stages. They were actively engaged in every process, be it collecting the data, connecting with the contributors or procuring relevant information. The team has been an ardent support to the editorial, designing and production team. Their endless efforts to recruit the best for this project, has resulted in the accomplishment of this book. They are a veteran in the field of academics and their pool of knowledge is as vast as their experience in printing. Their expertise and guidance has proved useful at every step. Their uncompromising quality standards have made this book an exceptional effort. Their encouragement from time to time has been an inspiration for everyone.

The publisher and the editorial board hope that this book will prove to be a valuable piece of knowledge for researchers, students, practitioners and scholars across the globe.

List of Contributors

Elsa Garmire
Dartmouth College, USA

Roberto-Carlos Fernández-Hernández, Lis Tamayo-Rivera, Alicia Oliver and Jorge-Alejandro Reyes-Esqueda
Institute of Physics, National Autonomous University of Mexico, Circuit for Scientific Research S/N, University City, Mexico City, Mexico

Israel Rocha-Mendoza, Raúl Rangel-Rojo
Department of Optics, Center for Scientific Research and Education Superior de Ensenada, Ensenada, Mexico

M. A. Ferrara, I. Rendina and L. Sirleto
National Research Council-Institute for Microelectronics and Microsystems, Napoli, Italy

Fabio Antonio Bovino and Maurizio Giardina
Quantum Optics Lab Selex-Sistemi Integrati, Genova, Italy

Maria Cristina Larciprete and Concita Sibilia
Department of Basic and Applied Sciences in Engineering, Sapienza University, Rome, Italy

G. Váró
Institute of Biophysics, Biological Research Center, Hungarian Academy of Sciences, Szeged, Hungary

C. Gergely
Montpellier University, Charles Coulomb Laboratory UMR 5221, Montpellier, France

D. Udaya Kumar, M. G. Murali and A. V. Adhikari
Department of Chemistry, National Institute of Technology Karnataka, Surathkal, P. O. Srinivasnagar, India

A. John Kiran
Department of Inorganic and Physical Chemistry,Indian Institute of Science, Bangalore, India

Valery Serov
University of Oulu, Finland

Hans Werner Schürmann
University of Osnabrück, Germany

D. Y. Wang and S. Li
School of Materials Science and Engineering, The University of New South Wales, Sydney, Australia

Carlos G. Avendaño and Ismael Molina
Autonomous University of Mexico City, G. A. Madero, Mexico D. F. Mexico

J. Adrian Reyes
Institute of Physics, National Autonomous University of Mexico, Mexico D. F., Mexico

Kazuya Ogawa
Nara Institute of Science and Technology, Interdisciplinary Graduate School of Medical and Engineering,
University of Yamanashi, Japan

Yoshiaki Kobuke
Institute of Advanced Energy, Kyoto University, Japan

Printed in the USA
CPSIA information can be obtained
at www.ICGtesting.com
JSHW011421221024
72173JS00004B/626